Flying Buttresses, Entropy, and O-Rings

The World of an Engineer

Harvard University Press   Cambridge, Massachusetts   London, England

Flying Buttresses, Entropy, and O-Rings

The World of an Engineer

James L. Adams

First Harvard University Press paperback edition, 1993

Library of Congress Cataloging-in-Publication Data

Adams, James L.
   Flying buttresses, entropy, and O-rings : the world of an engineer
   / James L. Adams.
      p.   cm.
      Includes bibliographical references and index.
      ISBN 0-674-30688-0 (cloth)
      ISBN 0-674-30689-9 (paper)
      1. Engineering—Popular works. I. Title.
TA148.A32 1991
620—dc20

91-11023
CIP

Contents

Introduction   1

1   A Brief History of Technology   5
The Underpinnings

2   Beyond the Calculator   31
The Complexity of Engineering

3   The Origin of Problems   58
The Pushes and Pulls

4   Design and Invention   78
The Concept

5   Mathematics   106
The Numerical Mystique

6   Science and Research   129
The Basics

7   Development, Test, and Failure   150
The Proof of the Pudding

8   Manufacturing and Assembly   177
The Critical Art

9   Money and Business   204
The Grease

10   Regulation   221
The Painful Inevitability

11   Thinking Technology   239
The Challenge for the Future

Sources and Suggested Readings   249
Acknowledgments   257
Illustration Credits   258
Index   260

Flying Buttresses, Entropy, and O-Rings

The World of an Engineer

This is a book about engineering. Since it examines engineering in a broad context, it is to some extent also a book about technology. Engineering and technology are not well understood by most people. The activities included are complex and varied, involving mathematics, physical sciences, and other areas of expertise that have evolved their own specialized languages and do not always communicate well, even with one another. Many people involved in engineering and technology are so caught up in their particular activity that they have not developed a good sense of the whole. And very few of those who have grasped the bigger picture spend much energy trying to communicate this sense to others. As an engineer I am often embarrassed by the relatively small number of readable general books on engineering, as opposed to the large number on the physical sciences and mathematics. That is one of the reasons I wrote this book.

But why should anyone outside the field desire to better understand engineering? First of all, engineering and technology are among the major long-time activities of humans and reflect all of our wonders, traumas, and flaws; not to understand them is to lose a tremendous opportunity for intellectual stimulation. Second, they are inseparable from the way we live. We cannot avoid engineering and technology, and a knowledge of them stands us in good stead whether we earn our paychecks by designing computers, haggle with the mechanic who will repair our automobile, or merely argue about the potential of genetic engineering with our friends at a dinner party. Finally, societies such as ours have a major investment in engineering and technology and are capable of doing great good or ill with them. We have reached a point where we can destroy ourselves with our technology in an uncomfortably short time. A society as technologically based as ours cannot remain functionally democratic, in the sense of making informed choices about our future, without an electorate and leaders who understand technology and the engineering process that is its core.

This book is intended for people who are not engineers—for general readers; for people managing, working for, or living with engineers; and for students either seeking a liberal education or considering engineering as a field. However, I hope that those in the profession will also find it of value. It is possible to become quite involved in engineering without acquiring a broad view of it, and this book should provoke thoughts about the context of anyone's technical work. If you are a professional engineer reading it for perspective, you are unlikely to discover that my viewpoints and perspectives exactly match yours, since engineering is a complex and multifaceted activity and since engineers tend to operate in a world centered around their particular field of expertise. However, there will be some match, and if the mismatch is severe, perhaps it will encourage you to write your own book on this topic, to help address the biases of the material in mine.

If you are thinking of becoming an engineer, this book will tell you something about the profession, reassure you that engineering is not all calculation, and perhaps worry you that engineers must live with more un-certainty than you thought. If you live with or work around engineers, this book should add to your understanding of them. If you are reading it purely for interest, I hope you not only will find it instructive but will also become convinced that engineering and technology are a complex and integral part of life. Technology is like our gastrointestinal tract: We can't live without it, so we might as well understand it better, because upset stomachs and the potential for stomach cancer are mixed right in with the ability to convert plants and animals into the chemicals we need to live and with the good feelings that result from a wonderful meal. Many people like to think that the uncomfortable aspects of technology can be separated from the obviously beneficial; that failures can be halted while experimentation continues; that "military" tech-nology can be separated from "civilian" technology; and that infinite amounts of cheap energy can be obtained rapidly without pollution. Unfortunately, the good and the bad seem to be discouragingly intertwined, and all are part of the life we have chosen. We can deal with the situation better if we better understand it.

This book is biased toward my own background and experience, simply because I am more comfortable talking about the things I know. Therefore, you will find more things mechanical than electrical or chemical. Although the theory, the words, and to some extent the interests of people who design freeway interchanges are different from those of people who manufacture tractors or grow near-perfect crystals for integrated circuits, there is consid-erable similarity in the process by which all engineering decisions are made.

It fascinates me to watch students worry so much about which field of engineering to study, since once they leave school they will not only cross disciplines but may change them more than once. This is not a book about gee-whiz devices and software—supercomputers, laser-guided weapons, transportation systems based on superconduction, and integrated circuits containing yet again ten times as many devices. We want to focus on the *process* of engineering here, rather than its glamorous products, which have an unfortunate tendency to become less impressive over time. Ten years ago, many magazines were featuring photographs of integrated circuits on their covers. Now these chips are routinely used in most devices, and the competition is waged as much over price as uniqueness of design.

I occasionally visit an aerospace company for whom I worked as an engineer 25 years ago. The tools used by the engineers—such as computer software for calculation and simulation—and the available materials and components certainly have changed. The engineers have also learned a lot about making their products more reliable, efficient, and sophisticated in this time. However, the overall character of the problems and approaches to solving them have changed very little. Carefully designed and constructed mechanisms still fail. Electrical and mechanical engineers still wage war about antenna size on spacecraft, attempting to impress each other with quantitative arguments based upon different value systems (high communication-system performance versus low weight and reliable deployment). Congress, the funding agency, and the media are still constantly on their backs. Late nights and tedious work are still required. Also, of course, there is the thrill of working at the edge of human ability on things that have never been done. This type of material is the primary focus of this book.

I begin with a brief overview of the history of technology and the engineering profession. Technology has been with us a long time, but only recently did it take on its present character. A bit of thinking about our history allows us to understand a great deal about where we are at present. Chapter 2 discusses the nature of engineering and introduces us to its complexity and the difficulty of making generalizations about it. Chapter 3 is concerned with why engineers focus on certain problems rather than others. The remaining chapters of the book each discuss a major aspect of engineering. Chapter 4 is concerned with design and invention—the part of engineering that produces new ideas. Chapter 5 has to do with mathematics, the discipline that allows modern engineers to build such sophisticated products and causes technology to become so mystical to the lay public. Chapter 6 discusses science and research, the source of some (but certainly not all) directions of technology. Chapter 7 has to do with the mysterious and important processes of experiment and failure. Newly designed products seldom live up to expectations. They must be "debugged" and still may sometimes fail. Chapter 8 is concerned with manufacturing and assembly, the processes by which the products of technology reach their final form. These essential processes were taken for granted for a number of years but have again caught the attention of U.S. industry, largely because other countries seem to do them better than we do. Chapter 9 discusses business and money. Technology requires the latter, which often becomes available through the former. To divorce engineering from business and money is to engage in fantasy. Chapter 10 grapples with the controversial topic of technological regulation in a free-market economy. Our technology has reached a level of maturity where it clearly must be controlled for the common welfare. Despite the loud outcries of business, the marketplace does not seem to be adequate for the job of control. We are presently struggling, through legislation and the judicial process, to learn more about how to control our technology in ways that are consistent with our economic and political values. This chapter is concerned with some of the whys, as well as some of the hows. And finally, Chapter 11 speculates on a few of the directions technology might take in the coming decades.

1

A Brief History of Technology

The Underpinnings

What are engineering and technology? How long have they been around? What do they have to do with mathematics and science? Who does them? What can they accomplish, and why do they seem to result in so much damage as well as good? To what extent does our society control them, and to what extent do they determine our lives?

These questions are becoming critical as our technology becomes ever more capable of transforming our lives for better *and* worse. Thanks to technology, we can take better care of ourselves than at any other time in our history. But we also have technology to thank for an unprecedented ability to exploit others and destroy ourselves. As an example, think about the so-called developing countries, who face an ever-increasing economic handicap based on relative technological capability. These days one needs modern technology in order to be successful in international economic competition. However, over time technology has become increasingly capital-intensive. Since the developing countries do not have sufficient capital, they become locations for factories and assembly plants that draw on cheap labor to produce products for foreign companies. Although manufacturing and assembly increase income and technological ability in the developing countries, they do not usually bring these nations to the point where they can compete with the developed countries in technological innovation. To further complicate matters, modern technology is substituting machines for human labor through automation based on digital computers, thereby decreasing the importance of cheap labor. As automation proceeds, the incentive even to manufacture and assemble in developing countries may decrease. Japanese automobile companies are now manufacturing in the United States, hardly a developing country. How can Sri Lanka or Ethiopia break into the automobile industry? The gap between the technological haves and have-nots continues to widen and is a long-term problem for us all.

Another example of the dangers of technology is of course the potential effect of the nuclear warheads scattered about the earth. At the time of this writing, tension in the United States about the likelihood of nuclear war has lessened, in response to political developments in the Soviet Union. However, those directly involved in arms control have not relaxed, because the warheads still exist and uncertainties in the Soviet Union about which factions are in political control have increased. In addition, other nations will likely continue to increase their missile-boosted nuclear capabilities. Our euphoria over the end of the Cold War will pass, and the threat of madmen using nuclear weapons will continue to weigh on us.

Some questions about the role of engineering in modern life can be answered by looking at history. Technology has been around a long time, but it has changed in character a great deal. The extent of these changes has not been anticipated, and this will likely be true in the future. History does not tell us when the next developments will occur and how drastic they will be, but it does indicate the nature of the game. Technology as we know it today is new in the time scale of human experience, and gaining a sense of its evolution is a good beginning for discussing the nature of engineering.

For most of the history of the earth, there has been no technology because there have been no humans. The earth is approximately 5 billion years old. Humans have existed for only 2½ million years, or about 1/2,000 of the life of the earth. Through the majority of this time, little approximating modern technology has existed. Tools have been found that are 2 million years old, so we can assume that humans have always used them. However, early tools were either found (a conveniently sized rock with a pointed end) or fabricated from a natural object by scraping it or banging on it with another object. Probably individuals made their own tools—no R&D labs, no assembly lines, no unions and leveraged buyouts. In fact, for most of our 2½ million years, the process we have used to fabricate products has been individual craft that has had little resemblance to modern technology. For 1½ million years, humans did not even have fire, much less sophisticated implements. They huddled in the cold, eating raw food, and their tools were modified sticks and rocks.

After we learned to use fire, we became a little more technically sophisticated. Traces of fire-hardened wooden spears have been found that date from 400,000 years ago. Specialized stone handaxes were in use 200,000 years ago. Stone hand tools, including impressive forms and carvings that must have been produced by other well-made tools, have been found dating from 50,000 to 20,000 years ago. Bows and arrows were invented about 15,000 years ago. The sophistication of some of these products implies that there were probably "professional" toolmakers, who fabricated tools for others in return for compensation. The forms of some of these tools and the decorations applied to them also tells us that there was art in the craft. Utilitarian objects, such as spear throwers in use 15,000 years ago, were often carved with extraordinary and beautiful animal figures and other symbols.

Modern technology, like ancient technology, requires ingenuity, craft, and art. The word *technology* is derived from the Greek words *techne*, which means art or skill, and *logia*, meaning science or study. The word *engineer* is from the Latin word *ingeniatorem*, meaning one who is ingenious at devising. In most languages, this derivation is clear. Unfortunately, in English we have confused the issue by taking the word "engine" from the same root. Engineers in English-speaking countries therefore drive trains, run power plants, and help fly airplanes as well as being ingenious at devising. However, being ingenious at devising is a characteristic of humans, and the engineer is expected to be a specialist at it. We will look more carefully at this requirement in Chapter 4.

About 10,000 years ago, during the neolithic period, a major change occurred in the lifestyle of humans. For the 2,490,000 years before that, we had been hunter–gatherers, but rather suddenly many groups of people changed careers and became farmers. They began to domesticate animals and live in villages and towns. As more and more people began to stay in one place and live in larger groupings, the sophistication of "technology" rapidly grew. It was still based on ingenuity, craft, and art, but, for instance, now more effort was spent on the construction of buildings, since they were to be used for a reasonable period of time rather than abandoned when the seasons changed and the hunting grounds moved. The new agricultural lifestyle required better food preservation techniques and storage—thus, the discovery of the utility of pottery, which has been found in Japan dating from 10,000 B.C. Farmers also had more time to devote to manufacture; textiles, probably woven on a simple loom, have been found from 6000 B.C. Settled people developed trade; and writing, which was first used for keeping track of taxes and trade, was in use in 3600 B.C. in Sumeria. Standard weights and measures were in use in Egypt in 3000 B.C.

Complex devices that would have been too cumbersome for a nomadic existence were soon invented. Wheels and boats were in use in the Nile–Euphrates valley by 3500 B.C. Plows were being used in Babylon in 3000 B.C., and potters' wheels appeared about the same time. Paved roads and sewage systems appeared in Crete in 2000 B.C. During this period we also began to use metals. The first known use of copper was in Turkey in 6400 B.C., and by 3000 B.C. mining and metallurgy were being pursued on an appreciable scale in many settled areas.

Once people began to stay put, more resources became available for large construction projects. The most impressive were the building enterprises of the Egyptians. They were royal works, under the supervision of experts in planning and construction who were on the staff of the king. These experts acted as engineers, architects, and project managers and attained high status in the kingdom. Imhotep, sometimes called the father of masonry construction, was a builder of tombs. Khufu-onekh was the builder of The Great Pyramid. Uni constructed canals and other hydraulic works. Ineni and Senmut were obelisk experts. After 5,000 years we still know of them. All of them were close to the monarch and had great influence not only in building projects but in other affairs of state as well. As to their engineering role, they were probably the bosses rather than the people actually doing the detailed technical work, since such names seem to endure better than the names of those lower in the ranks.[1]

The biggest problem facing these early engineers was time: Kings and pharoahs wanted to see results within their lifetime. Since the work was done by slaves, the welfare of the workforce was freely sacrificed in the interest of saving time. Not only did slaves receive no salaries, but they paid with life and limb for these symbols of absolute monarchy. Motivation of workers was through the whip, and safety regulations and medical insurance were not items of concern for the managers. These early construction managers created their work through untold human agony, but one must admit that they were rather impressive at the unprecedented organization and direction of huge numbers of people. The Great Pyramid, for example, was said by Herodotus to require the labor of 100,000 men for 20 years to cut, transport, and position the 2¼ million 2½-ton blocks. R. Engelbach has described the quarrying of obelisks, in particular an unfinished one in Aswan.[2] This obelisk was being cut from solid rock in one piece and was 137 feet long, 13 feet 9 inches thick at the base, and—had it been extracted—would have weighed 1,168 tons. He conjectured that relays of workers, standing two feet apart, took turns pounding the granite with hand-sized rocks of harder material to make the channels around the obelisk. He estimated that freeing the obelisk from the bedrock would have required approximately seven months, and erection would have been done by moving it up a long earth incline and then lowering the base end by gradually removing the earth—tasks that obviously required coordination.

By the time of the Egyptian empire, engineers (or at least their early equivalents) were still trained as craftsmen through apprenticeship and expected to be ingenious at devising. But they now had access to sufficient resources to create highly visible works that garnered greater social rewards. The age of the manager of technology had arrived. The bureaucratic processes involved in building pyramids were different from those that come into play when building a new airport in Denver, since these early civilizations were not democracies. Materials and equipment were primitive by modern standards, and skills such as labor negotiation were not needed, since slaves were not encouraged to form unions. However, technology had reached the point where structures could be produced on a scale that dwarfed the productive ability of a single person.

The Egyptian builders were also taking advantage of quantitative and mathematical relationships, another characteristic of modern engineering. The Great Pyramid, for instance, was built upon uneven ground and is oriented to the points of the compass within an accuracy of one in one thousand. This achievement required reasonably sophisticated surveying.

At about the same time as the pyramids were being built in Egypt (3000 to 2500 B.C.), impressive accomplishments were taking place in Mesopotamia. In 3000 B.C. the Mesopotamians were using the true arch. About 2000 B.C. the Code of Hammurabi, a complex system of law, was written. Since it included rules about construction, it is sometimes described as the first building code. It imposed the following rather interesting stipulations on the early builder:[3]

If a builder has built a house for a man and has not made his work sound, and the house which he has built has fallen down and so caused the death of the householder, that builder shall be put to death.

If it causes the death of the householder's son, they shall put that builder's son to death.

If it causes the death of the householder's slave, he shall give slave for slave to the householder.

If it destroys property, he shall replace anything that it has destroyed; and, because he has not made sound the house which he has built and it has fallen down, he shall rebuild the house which has fallen down from his own property.

Many years before malpractice insurance, these early engineers in Mesopotamia were facing the harbingers of liability law. What it says to us is that builders were considered specialists whose products were sophisticated enough that society saw fit to regulate them to ensure the desired quality.

As empires developed, so did warfare, and warfare has always been one of the chief motivators of technology. The Assyrians showed impressive prowess in military engineering, making use of fortified towns, battering rams, armored wagons, pontoon bridges, and especially iron weapons. Sargon II was king of the empire from 722 to 705 B.C., and a single room of his palace when exacavated was found to contain 200 tons of iron weapons.[4]

Consider the Greek civilization, of so much importance to us "Westerners." Rightly or wrongly, we tend to identify much more closely with the culture of the ancient Greeks than with earlier civilizations such as Egypt, Mesopotamia, or Assyria. What was the status of technology in their time? The Greeks depended upon skilled technology (exports of metal and metalwork, cloth and pottery, oil and wine, and construction of roads, buildings, and even water-powered mills). However, the more prestigious Greek citizens were theoreticians, involved in philosophy, government, mathematics, and other nonmaterial pursuits. They were unprecedented in their emphasis on thinking, introspection, and intellectual invention and had relatively little interest in reducing their concepts to practice. Natural philosophy, sometimes referred to as the precursor of science, began in Greece about 1000 B.C., with thinking about the properties of numbers, the properties of triangles, and the nature of fire.[5] Aristotle (384–322 B.C.) was perhaps the dominant intellectual influence of the classical Greek period in this respect, but he referred to the earlier conjectures of Thales (seventh century B.C.). The school of Pythagoras (582–500 B.C.) had already laid the foundations for geometry, followed by the work of Euclid (approximately 300 B.C.). Democritus (470 B.C.) speculated on an atomic base for matter, and Galen (130–201 A.D.) developed medical theories that were to last for a thousand years. However, these early thinkers eschewed experiment and therefore had no way of calibrating their theories. Science as we know it did not exist. And if it had, it is not clear that it would have been applied to problems of trade and technology. (This will be discussed in greater depth in Chapter 6.)

The Greeks were similarly inventive in mechanical technology. Archimedes invented compound pulleys, hydraulic screws, burning mirrors, and various engines of war. In keeping with the Grecian status system, however, he considered his inventions to be intellectual exercises, and he concealed any pride he might have gotten from their utilization. At one point in his life he publicly denied the importance of machines of war that he had designed, not because he was against war but rather because he did not want to be seen as a mere builder of physical devices.

Hero of Alexandria invented devices for raising weights with derricks, a right-angle sighting device called the dioptra, presses, fountains, a rotary steam turbine, and a water pump conceptually identical to one used in nineteenth-century fire engines. But like Archimedes, Hero did this primarily as a mental exercise, and when the devices were built they were used purely for show and for such functions as automatically opening temple doors. They were not used to affect the standard of living of the people living in Greece. They were curiosities designed by a philosopher.

Greek technology also made use of quantitative relationships. For instance, the following is a formula used to define the caliber of a Greek catapult:[6]

Multiply by 100 the weight—in minas—of the stone missiles. Take the cube root of the product and add 1/10 of the number thus produced. This gives the caliber in daktyls.*

*The Greek dactyl was approximately ¾ inch, or 19.3 mm. The minas was just under a pound, or 436 grams.

or d = 1.1 × cube root of (100 × weight in minas)

For example, if the weight of stones is 80 minas

$$80 \times 100 = 8000$$
$$\text{cube root of } 8000 = 20$$
$$1.1 \times 20 = 22$$

Therefore the calibre will measure 22 dactyls. If the cube root is not an exact number it is rounded off by the addition of one tenth.

This may seem similar to relationships that modern engineers use, but it is a far cry from present practice. The relationship was empirically determined rather than based on an understanding of physical principles. But it is a precursor that shows a desire for the type of understanding which would enable the technologists of the time to optimize their works based on the experience of others.

The attitude of the influential Greeks toward technology is important in understanding attitudes in modern Western countries. In Greece, the higher classes were not involved in trade or technology. The activities associated with technology were considered to be more properly done by slaves and foreigners, while citizens involved themselves in more disembodied mental activities. This attitude has been passed down to us through Western intellectual tradition and continues to influence Western societies, even though the nature of technology has changed radically. It is partially responsible for the intellectual stereotyping associated with engineering. After all, Oxford and Harvard do not teach engineering, and the engineers our society admires have often achieved wealth and social power by organizing others (managing), or by leaving engineering in order to enter the area of politics, philanthropy, teaching, or philosophy—pursuits much more compatible with the ideal of ancient Greece than choosing the proper ball bearing or worrying about defects in the coating of a reflective surface.

By contrast with the Greeks, the Romans made their mark as technological implementors, rather than inventors. They had access to abundant labor (slaves again), materials, and simple structural principles, and they prided themselves on their building. Many emperors were known by the construction that took place during their reign. Claudius (10 B.C.–54 A.D.), for instance, completed the major harbor works at the port of Ostia that allowed Rome to receive the grain imports that it required. Tourists still visit Trajan's column in Rome. Like the Egyptians, the Romans became extremely competent at the management of large-scale technological projects and afforded their engineers, or at least their project managers, high status compared with the Greeks. As an example, Frontinus, who was responsible for the Roman water supply from 97 A.D. to 104 A.D., was honored twice with the consulship and referred to as a colleague of the emperor. He not only supervised some 250 miles of conduit and 700 workers but wrote a book entitled *De Aquis* on the water system and his own approach to technology.

Also to their credit as engineers, the Romans placed great emphasis on standardization and organization. By their emphasis on repetitive use of traditional materials and forms, they reduced their technology to simple empirical relationships that could be transmitted from engineer to engineer. The first engineering textbook we know about was *De Architectura*, written about 15 B.C. apparently by Marcus Vitruvius Pollio and later used by the builders of the Renaissance. This was a book of information that Roman builders should know. It was based on experience and was clearly for the use of people who did not worry about the physical rationale of its many rules of thumb. These dealt not only with details on surveying, technique, and devices but also with the education and values of the builder.

At the end of the Roman Empire, therefore, Western technology had a base in ingenuity, craft, art, empirical relationships, and some quantitative reasoning. We had learned to organize large numbers of people (usually slaves) and conduct extensive building projects (usually for kings and emperors). Natural philosophy had set the stage for science, and we had become more sophisticated about mechanical devices, although they were much less important to early civilizations than civil and military works. Technology had brought us increased sophistication in war, transportation, agriculture, buildings, and personal possessions such as clothing, housewares, and ornaments. However, it was still based on the energy of humans and animals and had not yet produced precision mechanisms nor a means of communicating beyond the range of couriers and signals that could be seen by the eye. It made only slight use of mathematics and none at all of science.

During the Roman Empire and up to the medieval period in Europe a tremendous amount of technological activity had occurred in Asia and Arabia. After the fall of the Roman Empire, technology was generally more advanced in the Near East (what we now call the Middle East) and Far East than it was in Europe. Greek and Roman knowledge were augmented and new inventions and principles added. In the medieval period, this technology was imported to Europe from the East through overland trading and political/religious movements such as the crusades. As examples, iron casting, the wheelbarrow, the stern post rudder, the crossbow, gunpowder, paper, silk-working machinery, and printing with wood and metal blocks arrived in Europe during the twelfth and thirteenth centuries, probably from China, where they had all been used for hundreds of years.[7]

Many of us were taught that Europe entered the "dark ages" between the fall of Rome and the medieval period, and that even the medieval period was not so intellectually exciting. Historians are now revising this opinion, especially as far as technology is concerned. After the fall of Rome, Europe became dominated by small city-states and more nomadic cultures, which did not leave as splendid a supply of impressive artifacts and written records as did the earlier empires. During this time, though, technological development continued. It is true that some technological capability disappeared with the Empire; the Roman art of making concrete, for instance, seems to have vanished for about a thousand years. However, the foundations of modern mechanical power were slowly being laid through expanded use of water. Water wheels and crude gearing were known by the Greeks, and the Romans had constructed very large mills. Flour mills using these principles became increasingly common after the fall of Rome. According to the Domesday book, an inventory of England prepared by order of William the Conquerer, there were nearly 6,000 grain mills in operation in 1085. These mills used water power to turn millstones to crush the grain. In the ninth century the cam (also known to the Greeks) was added. A cam is a rotating eccentric shape that allows rotary motion to be translated to linear motion. This development allowed the addition of hammer mills, saw mills, and pumps to the inventory of powered devices.

The stirrup (imported from Central Asia through Byzantium to Europe in the eighth century A.D.) allowed the rider to use heavy weapons including the lance and therefore was instrumental in the development of knighthood, chivalry, and the feudal system. The heavy plough (in use by the Slavs in the sixth century) led to radically increased food production through its ability to handle the rich, moist alluvial soil of much of Europe. The horse collar (invented in the ninth century) and the horseshoe allowed much more power to be applied to the plow. Before these inventions the ox had been a superior work animal to the horse because of its tougher hooves and because the yoke collar in use had choked the horse. If successfully harnessed, however, the horse can deliver 50 percent more work in a given time than an ox and has greater endurance. The horseshoe, the horse collar, the heavy plow, and crop rotation (begun in the eighth century) resulted in an agricultural revolution that led to a greatly improved standard of living.[8]

Thus, in the year 1000 Western Europe had developed some water and wind power, but it remained dependent on human and animal energy. It had few large cities and little trade. Its economy was based on subsistence farming, and there was still little technology as we presently define it in Europe after 2,499/2,500 of human history (4,999,999/5,000,000 of the age of the earth).

During the medieval period which followed, many developments demonstrated that technology was very much alive in the West.[9] Among the more dramatic engineering feats were the Gothic cathedrals, many begun in the thirteenth century, which showed not only that the West had retained the ability to concentrate labor and money but also that sophisticated advances had occurred in technological understanding. The structural challenge of the Gothic cathedral was not only to obtain great spaces and soaring ceilings but also to maximize the amount of glass in the walls in order to obtain an interior impression of light and airiness. Much use of the arch was made to accomplish this. Rib vaulting simultaneously provided adequate strength while reducing overall weight. Flying buttressing offset lateral loads at the points where the arches met the vertical piers (see figure, next page). The flying buttress, so evident in cathedrals such as Notre Dame, adds to the lightness of the structure by soaring through space to do its duty, rather than being solidly built up from the ground. Considerable technological ability is necessary to design such structural elements, as well as the machinery and forms necessary to build and erect them.

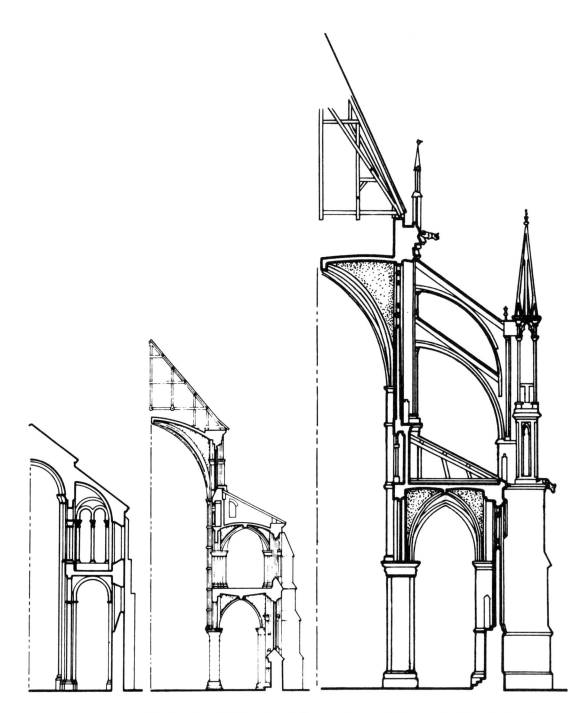

Sections through Sainte-Foy-de-Conques, Laon, and Reims cathedrals, showing the evolution of the flying buttress.

Modern structural analysis applied to some of these cathedrals has shown that their design is surprisingly sophisticated. For example, the chief goal of masonry construction is to avoid placing any of the material in tension, because masonry is very strong in compression but weak if you pull on it. Modern buildings solve this problem by reinforcing the masonry with steel, which is very strong in tension. However, there was an unfortunate shortage of low-cost reinforcing bar in the thirteenth century, so cathedrals were made of stone and concrete alone. We now know that many of the pinnacles and statues on buttresses, which were formerly thought to be purely ornamental, actually serve to offset forces tending to pull the structure apart. In technical jargon, they change tensile stresses to compressive stresses.[10] These buildings were made by skilled technicians who had acquired enough knowledge through experience to achieve structural efficiency that today is producible only by highly educated engineers using sophisticated theory backed up by powerful computers.

Rapid advances also occurred in arms, armor, and siege weapons during the Middle Ages. The armored knight could overwhelm the foot soldier but was in turn at the mercy of the pike and the crossbow. Perhaps most important, reliance on water and wind power to augment physical output sharply increased. The typical abbeys of the twelfth century, for instance, used water in very sophisticated ways. The water system at Clairvaux Abbey, as an example, provided power for corn mills, crushing mills, a fulling mill, and a tannery, as well as supplying the kitchen and washing places and flushing the latrine.

During the early medieval period, the manufacture and distribution of goods in Europe were heavily influenced by guilds. The merchants' guild—an association of traders involved in international commerce—sought not only to establish monopolies in trade routes for its members but also to influence the manufacture of products. Crafts guilds, which represented specialized artisans who produced goods, succeeded in improving the quality of merchandise by establishing production standards and increasing the rigor of apprenticeship programs.

The period 1250–1350 in Europe is sometimes referred to as the age of invention (especially mechanical invention). The elements were in place to allow an unprecedented level of ingenious devising (see figure). This was the time when spinning and weaving devices, weight-driven clocks, and mechanized universe models were being rapidly developed. Great improvement of the firearm (actually first invented about 1050) made obsolete many ancient and traditional aspects of warfare, although it took the military many years to accept it. During this period there was also tremendous interest in innovation and in quasi-technological topics such as perspective. Our forebears were becoming more knowledgeable but still lacked many elements of modern technology.

During the Renaissance a great flurry of intellectual and economic activity took place that was to affect technology. Such influential people as Leonardo da Vinci (1452–1519) combined a love of invention and understanding with great respect for mathematics. Competition among wealthy city-states resulted in works such as Brunelleschi's (1377–1466) great dome on the Florence cathedral. The pace of creative activity was heightened by a new mobility among scholars and craftsmen due to competition between patrons and by improvements in communication brought about by printing. The first major treatise on architecture in the West since *De Architectura* was written by Leon Battista Alberti in 1452. Because reproduction was now cheaper and faster, *De re Aedificatoria* was quickly and widely influential, and builders who could read it had a professional advantage over those who could not. In a sense, the age of publish (or at least read) or perish had begun.

The emphasis on studying ancient "science" in the Renaissance (which was based on the thinking of men such as Aristotle, Ptolemy, and Galen) eventually led a few talented people to develop new paradigms to explain some of the fallacies they uncovered in the old theories. Aristotelian theory, for instance, had concluded that a heavy weight would fall faster than a lighter one. Even in classical times, this had bothered some people. If two equal weights were dropped, Aristotelian theory would say that they would fall faster if fastened together. Aristotle was also violently opposed to the concept of a vacuum; yet vacuums would soon be not only proven but would become the subject of much experimentation and application. The thinking of those skeptical of classical natural philosophy resulted in scientific theories based more on mathematics and empirical observation than abstract thought—a direction

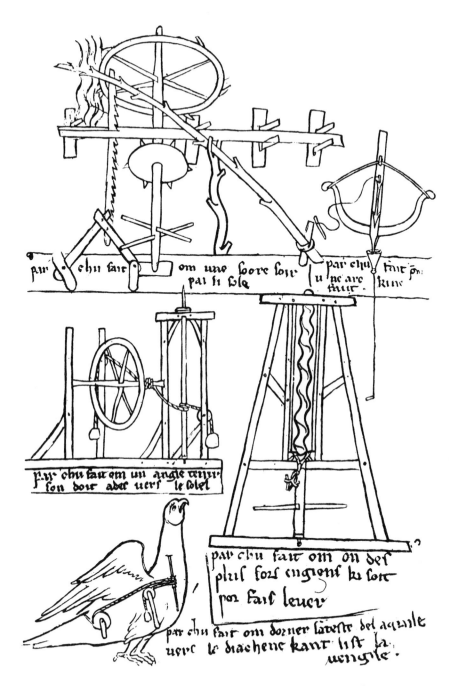

par cha fait om uno boore foŋ
par li bsle

par cha fait p̄
u ne arc
faiŧ.
kini

par cha fait om un angle œŋ̄
son doit aœr uers le bolel

par cha fait om on der
plus foŋ ꝯgionŧ k̄ fon
por fait leuer

par cha fait om dozner faueŧe œl aquile
uerc le brachene kant lift la
uengile.

A saw powered by a water wheel and returned by a sapling; a screw jack; an escape-
ment; and an articulated decorative eagle, all from the 1240 notebook of Villard de
Honnecourt.

that would eventually lead to greater interaction between science and technology.

This movement took place rather slowly at first, because, among other things, the Church was comfortable with the worldview formulated by the ancients and rediscovered and interpreted by Renaissance intellectuals. They had little use for the new mechanistic view of the universe that was slowly emerging, nor for cosmological models and notions of mankind's place in the universe that grew out of this kind of thinking. For example, the Church was quite content with the earth-centered model of the cosmos as elaborated by Ptolemy, because it put mankind at the center of a perfect universe. Ptolemy, an Alexandrian mathematician and astronomer of the second century A.D., had been highly influenced by Plato's notions of perfection, particularly the perfection of the circle. For Plato, anything as important as a planetary orbit should be perfect. Consequently, Ptolemy concluded that the planets moved in small, perfect circles, or epicycles, whose centers traveled in larger circular orbits centered on the earth.

The Polish astronomer Nicolaus Copernicus (1473–1543) was the first Renaissance thinker to explain the observed motions of the planets by hypothesizing a sun-centered system (although among the ancients, Aristarchus had produced such a model in the third century B.C.). But Copernicus clung to the circular planetary motions of Ptolemy; his innovation was merely to include earth among the planets that orbit the sun. Johannes Kepler (1571–1630) finally determined that the planetary orbits were elliptical by using very detailed astronomical information collected by Tycho Brahe (1546–1601). Platonic notions about the natural world were finally being supplanted by a more mechanistic worldview and a more empirical approach to scientific investigation. By the year 1700 René Descartes (1596–1650) and Isaac Newton (1642–1727) had shown conclusively that not only was the sun at the center of the solar system but that the physical universe beyond the earth obeys simple mechanical laws.

The activities of the sixteenth and seventeenth centuries are sometimes referred to as *the* Scientific Revolution, though this was neither the first nor the last revolution in scientific thinking. (One is probably occurring at present.) However, during this time there was, undeniably, an impressive amount of intellectual activity, which led to scientific attitudes, methodologies, and theories that caused science to converge with technology. During this time Galileo Galilei (1564–1642) replaced Aristotle in physics, and scientists came to rely heavily on experiment to validate theory. Andreas Vesalius (1514–1564) and William Harvey (1578–1657) supplanted Galen in medicine, by performing detailed dissections and physiological studies which showed that the human body obeys the laws of nature as well as the will of the gods. William Gilbert (1540–1603) developed an understanding of magnetism during this period, and Agricola (the penname of George Bauer, 1494–1555) published his famous book on mining.

Many of these scientific discoveries depended upon advances in technology. Pendulum clocks allowed greater accuracy than water or sun clocks, and this accuracy in turn allowed much more precise measurements of rates and accelerations. The new emphasis on experiment required increasingly sophisticated technological equipment such as the air pumps and sealed chambers used by Giambaptista Porta (1536–1605), Evangelista Torricelli (1608–1647), and Otto von Guericke (1602–1686) to experiment with atmospheric pressure and vacuum. During the same period technology was ripe for the improved knowledge that could be gained by scientific experimentation and hypotheses. As an example, the great interest in steam as a source of energy led to a working steam engine in 1698. It was developed by Thomas Savery (1650–1713), lest you think that Watt invented the steam engine.

Although some time would pass before science actually "led" engineering, early hints of what was to come can be seen in the work of Galileo, who is commonly considered to be the inventor of engineering mechanics, which is the application of mathematics to predict mechanical behavior. Galileo sought to use his understanding of the natural world to derive relationships of use to the engineer. (This move toward engineering science will be discussed in more detail in Chapter 6.) Galileo and his contemporaries based their thinking on attempts to understand the phenomena involved, rather than relying solely on experience and cleverness. Newton (1642–1727) went one step further by demonstrating that the heavens obey the same laws as prosaic objects on the earth's surface. This approach was very different from that taken by the Romans and reflected in *De Architectura*. The Romans were interested in how rather than why and were very happy to leave the heavens to the gods.

One reason for this change in thinking was undoubtedly the Protestant Reformation and the increased intellectual freedom offered to intellectuals, who no longer bore the burden of ensuring that their theories were compatible with the teachings of the Church. Other reasons were the foment associated with the opening of the New World, changes in political structures and beliefs, and the realization that knowledge based on reality could be economically valuable.[11]

The eighteenth century is noted for great advances in mathematics. Mathematicians such as Joseph Lagrange (1736–1813), Pierre de Laplace (1749–1827), and Joseph Fourier (1768–1830) provided the mathematical underpinning for many developments in physics and technology. The invention of the compound microscope (which allowed much higher magnification through multiple lenses) made possible great strides in the life sciences. Unfortunately, the physical sciences were chasing various wild hares (phlogiston, caloric fluid) during this time, and chemistry had been hampered by long allegiance to the Aristotelian model (in which earth, air, fire, and water were the basic elements), which had resulted in the surprisingly rigorous but erroneous field of alchemy. However, in the eighteenth century such people as Antoine Lavoisier (1743–1794) and Amadeo Avogadro (1776–1856) put a rational base under chemistry, and in the beginning of the nineteenth century Sadi Carnot (1796–1832) developed thermodynamics. James Clerk Maxwell (1831–1879) was to follow with his brilliant insights into the relationship between electricity and magnetism, and science was in a position to truly contribute to technology.

Technology, of course, was not simply standing around waiting for science and mathematics to get their act together. In the sixteenth century, cloth and glassmaking grew in and around Venice, and the printing and instrument industries thrived. Great progress was made in the technology of mining. The seventeenth century saw construction of extensive canal systems in England and France, tremendous advances in shipbuilding, and continued experimentation with steam and vacuum.

The eighteenth century witnessed perhaps the greatest explosion of technology in history—the Industrial Revolution. It began with the production of machinery that enabled previously hand-done tasks to be performed by machine. The spinning frame, invented by Richard Arkwright (1732–1792) and John Kay (1733–1764), and patented in 1769, was a complex but ingenious arrangement of rollers and spindles powered by external energy (a water wheel). Hero's machines were a hobby; the cathedrals were built because of religious zeal; but here we have commercially significant labor-saving, economical technology.

These machines required power, and because of the availability of improved materials and manufacturing techniques, steam power became practical (see figure, page 24). Commercial steam power had been used for some time, but now James Watt (1736–1819) introduced external condensation, double action, the use of the governor, and improved precision that allowed a tripling of the efficiency of the early commercially successful atmospheric engines such as the one developed by Thomas Newcomen (1663–1729) and used to pump water from mines. A lathe designed by Henry Maudsley (1771–1831) about 1800 looks primitive by modern standards but has many of the elements of present lathes and the ability to handle steel in a sophisticated way (see figure, page 25).

The Industrial Revolution began with steam and steel and ended with transportation and communication. In 1852 Henri Giffard's (1825–1882) steam-driven airship flew 17 miles from Paris to Trappes, a significant technological accomplishment considering that thermodynamics had only recently been born. In 1876 Alexander Graham Bell (1847–1922) uttered his famous message, "Mr. Watson, come here—I want you," and the most critical tool used by modern teenagers was born.

During the Industrial Revolution, technology became essential to the economic health of Europe. As people moved to the cities, specialization increased, and it was no longer possible for the majority of people to fulfill their needs directly through subsistence farming. But the shift from farming or cottage industry to factory life was a difficult one for many people. A quote from a hosier of the time who worked in an English factory is illustrative: "I found the utmost distaste on the part of the men to any regular hours or regular habits . . . The men themselves were considerably dissatisfied, because they could not go in and out as they pleased, and have what holidays they pleased,

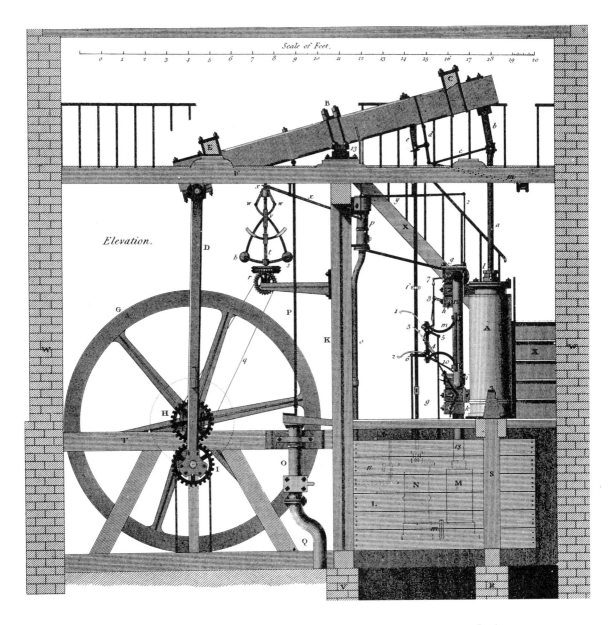

A commercial steam engine built in 1787 by the firm of Matthew Boulton and James Watt.

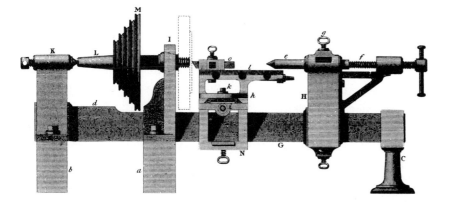

Henry Maudsley's screw-cutting lathe. Machines such as this allowed the production of precise, interchangeable parts.

and go on just as they had been used to do: and were subject, during after-hours, to the ill-natured observations of other workmen, to such an extent as completely to disgust them with the whole system, and I was obliged to break it up."[12] We tend to forget that the nine-to-five life is a recent invention; through most of history free people had more personal flexibility in their lives. The shift of Western societies toward coordinated work in an industrial setting produced major changes in the lives of most people and took many years to accomplish. The disparity in income and power between "workers" and "managers" was also apparent in a factory, and workers began to resent what they saw to be their exploitation by wealthy businessmen.

Of the many social protests against industrialization, the best known is probably the Luddite movement in the early nineteenth century, in which workmen tried to prevent the use of labor-saving machinery by destroying it and burning factories. The movement was supposedly named after Ned Lud, a man who broke up a stocking frame in the eighteenth century because of dissatisfaction with working conditions. These antitechnology protests could not compete at the time with the economic power of the revolution, the increase in material welfare, and the advantages of automating work that had previously been tedious, exhausting, and dangerous. But here and there people began to worry about the more disquieting effects of technology, such as filthy cities, vast accumulation of wealth by individuals, and intolerable working conditions.

The profession of engineering began to acquire its present characteristics during the Industrial Revolution. Famous engineers—especially those involved in business, such as John Smeaton (1724–1792) and James Watt—acquired significant income, prestige, and influence because of their technical abilities. For the first time, organized efforts were made to increase communication among engineers engaged in similar activities. One of the first professional societies to form was the British Institute of Civil Engineers in 1818, although even before this time there had been informal and formal get-togethers of engineers involved in similar work. At the time of its formation, "civil engineering" was the alternative to "military engineering" and therefore included all of those who were not directly involved in military works. The American Society of Civil Engineers was officially formed in 1852, after informal activities spanning some thirty years. These societies then proceeded to splinter into groups representing specific specialties or disciplines. In the United States in the 1880s, the American Society of Mechanical Engineers and the American Institute of Electrical Engineers were formed. In the 1890s the Society of Naval Architects and Marine Engineers, the American Society of Heating and Ventilating Engineers, and the American Railway Engineering Association were begun. New societies continued to proliferate, until by now there are over 130 national engineering and allied societies in the United States.

Formal education in engineering also did not begin until the Industrial Revolution. The first schools of engineering opened in France in 1757; the Ecole Polytechnic formed in 1794. In the United States, professional education in military engineering began at the Military Academy at West Point in 1817. Harvard, Yale, and Dartmouth adopted engineering curricula in the middle of the nineteenth century, although Harvard was later to drop its program, and separate technical schools were opened (Rensselaer Polytechnic Institute and the Massachusetts Institute of Technology). By 1870 there were some 70 schools teaching engineering in the United States; by 1990 the number had reached 250.[13]

From its earliest days, a major concern of technology has been war. The nineteenth century brought another generation of changes in military technology. During the American Civil War trains were used for quick transportation and the telegraph for rapid communication; rifled muskets, repeating rifles (the precursors of automatic weapons), and large amounts of artillery allowed unprecedented slaughter. Wars since then have proven that technology has rendered obsolete the type of warfare that occurred throughout most of history—noble charges, relatively low civilian casualties, and personal heroism. Although we children of the twentieth century still cling to our modern knights such as fighter pilots, and although our armed services talk of surgical strikes, all-out war between major powers is simply unacceptably messy, thanks to modern technology.

Another concern of early technology was building, which has also gone through tremendous change in this century. The design of buildings, like much of engineering, has benefited a great deal from the availability of modern computer-based analysis. Structures and associated mechanical systems can now be designed with a degree of sophistication unheard of even 30 years ago. New materials, such as structural adhesives, and new techniques, such as the replacement of nails with machine-driven screw fasteners, are revolutionizing construction. Modern elevators, mechanical systems, and communication techniques make new geometries possible. Construction sites are highly mechanized and computer controlled, and construction crews now rely on extensive and sophisticated communication equipment.

The chemical industry has come of age with the production of sophisticated polymers and pharmaceuticals through the use of modern catalysts and processes. But perhaps the most exciting area of technology at present is genetic engineering. This will be discussed in more detail later in the book, but the potential applications in chemical processes and the manufacture of commercial chemical products are enormous. Many major international concerns, such as health care, environmental pollution, availability of low-cost energy, and resource depletion, either directly or indirectly involve the chemical industry and chemical engineers.

The second half of the twentieth century has seen the flowering of the electronic and computer industries. As we will discuss later, the invention of the transistor, the integrated circuit, and the microprocessor has made revolutionary changes in technology and in our personal lives. Computers have also made a tremendous improvement in the professional activities of engi-

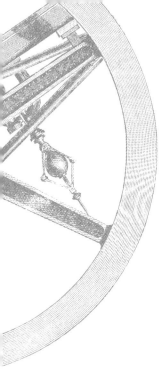

neers and in their ability to handle complicated problems. Before the modern digital computer, one could find rooms of people in aircraft companies spending each day performing detailed and repetitive calculations on mechanical calculators. Now computers are doing this work much more quickly, cheaply, and thoroughly and relieving engineers of hours of tedium. Even more important, computers allow an unprecedented complexity of analysis and simulation and the ability to deal with extremely complex physical and mathematical relationships. When I went to school, we spent a great deal of time learning to solve differential equations by hand, whether exactly or by numerical methods. Even after all of that time, we were practically limited in the types of equations we could solve. Now computers can do all of those calculations much better than we can, leaving us to do more important kinds of thinking.

Electronic technology such as modern telephone systems, duplicating and facsimile machinery, and computer networks and data bases have revolutionized the workplace of the engineer just as computer application has revolutionized the factory. As far as the products of electronic and computer technology are concerned, the list is overwhelming. As a minor example, I am writing this book on a personal computer that allows me to work many times faster than a typewriter; meanwhile, I am listening to music from a compact disc played through a sound system of extraordinary quality, while a telephone unit protects me by recording my incoming calls. These modern miracles cost surprisingly little, considering that they simply did not exist 30 years ago.

The rapid changes of the twentieth century have distorted our view of technology. We have become extremely focused upon what is called high technology—electronics (particularly information processing and communications), aerospace, genetic engineering, robotics, and other industries that employ highly educated people to invent products based directly on new scientific advances. The media bombard us not only with the real and potential benefits of these products but also with the problems associated with them. We cannot wait to obtain our facsimile machine, our cellular phone, and our camcorder, yet we worry about the effects of too much TV on our children, the cost of high technology in medicine, and the unknowns associated with genetic engineering.

In imagining the role of technology in the twenty-first century, we should not lose sight of the fact that many of our problems can be solved only with "low" technology. Repairing all of our damaged roads, bridges, rails, pipes, and other components of what is sometimes called "infrastructure"—a two-trillion-dollar project—will be accomplished with older technology. And when it comes to producing products with low enough prices and high enough quality to compete in the international market, we can look to computers and new materials and processes for help, but most of our production problems will not be solved with only higher technology. Our energy supply is another major problem, both because of the finite supply of petrochemicals and because of an atmosphere that is more vulnerable to pollution than we had thought. Perhaps some of the solution to our energy problems will come from increasingly efficient and cheap solar cells and superconducting devices and transmission, but most of our energy needs will continue to be met by improved generating plants, internal combustion engines, and other devices that have been around a long time. We will continue to want steel, concrete, lumber, fingernail files, pencils, and paper towels. Although the new technology is helping to produce all of these goods more efficiently, older technology is still crucial to the satisfaction of our needs. The improvements that can be made in the older technologies, though incremental compared with glamorous high-tech breakthroughs, are not trivial.

Despite the continued role of older technologies, as we look to the future we see no indication of a slow-down in technological ability, and in fact some who contemplate the next century think that the rate of change in technology will increase. Unfortunately, or fortunately, we are not infinitely flexible. I have for a long time been interested in change and creativity and do a good bit of consulting in companies that are based on changing technology and product evolution. People who work in companies like this pride themselves on their love of innovation and their ability to deal with rapid growth and product obsolescence. However, in actuality individuals, companies, and especially societies require traditions and operate by past experience. We do not instantly and painlessly incorporate new revolutions into our culture. In my opinion, the present rapid rate of technological change cannot effortlessly be incorporated into our community psyche, given that only 10,000 years ago we made our living hunting squirrels and gathering berries. We are not going to stop technological change because it is valuable to us in so many ways. But we are going to have to spend more effort understanding it and channeling it toward acknowledged goals.

Through the eighteenth century, social attitudes toward technology were generally positive. Not until the Industrial Revolution did significant negative feelings emerge as distribution of wealth and the nature of work changed. At present we have a mixed and complex set of attitudes toward technology. In the late 1960s and early 1970s in the United States the destructive potential of technology was vividly presented through media coverage of the Vietnam War, which was opposed by large numbers of people. During the late 1960s we also became conscious of large and long-term environmental problems. Widespread dissatisfaction with our frantic materialistic social values, as decried by the hippie counterculture, was followed by anxiety over inflation, the OPEC oil embargo, finite resources, international competition, and the possible ill effects of our third industrial revolution, brought about by computers and telecommunication. All of these events severely tried our love affair with technology.

At present, we seem to have swung toward a more positive but still realistic attitude. We purchase and use technology's products and look to it for the solution to many of our problems, while still having strong reservations about the downside. The increased application of science to technology is resulting in unprecedented uncertainties and unknowns, huge costs, and high visibility. We are justifiably worried about our technology-based economy, as international competition challenges our leadership and as the struggle over limited resources and their allocation increases. We suspect that we are destined to live in international political tension, and that we are becoming more and more vitally dependent on technology despite disagreements about which directions that technology should take.

Our society has to become "smarter" at managing its technology. We need to understand it better, and we need to encourage informed dialogue between those who are affected by it. One approach to increasing understanding is through history. This chapter has briefly mentioned some of the more important historical trends and happenings. Let us now move to a more detailed examination of the nature of modern engineering.

2

It would be nice to begin this chapter with a short and simple description of engineering. Unfortunately, there is none. You can tell that engineering is a complex subject by applying two short tests. First if you look in the dictionary, you will find definitions that at first glance seem reasonable but that, upon further thought, tell you nothing.

> The activities or the function of an engineer; esp: the art of managing engines. The science by which the properties of matter and the sources of energy in nature are made useful to man in structures, machines, and products (*Webster's Third New International Dictionary,* Springfield, MA: Merriam-Webster Inc., 1981).

The first phrase runs us in circles, and the second—"the art of managing engines"—would result in an expletive from most engineers I know. The next definition is grand, a bit pompous, and quite misleading. Whenever dictionaries have trouble, you can be sure the subject is complex.

A second test for complexity is the amount of stereotyping that exists about a topic. Here engineering is rich: engineering is the application of science to practical problems; engineers are nerds; engineers are thing-oriented, not people-oriented; engineers work on drafting boards in huge rooms; engineering is immoral, amoral, good, bad, boring, exciting. People like to simplify. If subjects are complex, they will stereotype in order to lessen the burden on their overworked minds.

Among the many reasons for this complexity, first of all we notice a bewildering cast of characters playing the game. Approximately 2 million people in the United States call themselves engineers. A measure of the complexity of the field is that different sources disagree on the number. By looking through the *1988 Statistical Abstract of the United States,* one is able to conclude that the number is somewhere between 1.3 million and 2.6 million engineers, a nice factor of two. They are not all nerds, although some of them are; some also have outstanding social skills. Some are unusually articulate; some are not. Some are poets, painters, musicians, and priests on the side. Some of them are murderers, drug dealers, and prostitutes. Jimmy Carter and Tom Landry were educated as engineers. So were Frank Schroeter and Elmer Floyd—excellent engineers unfamiliar to most readers.

In a study made some 20 years ago, Robert Perrucci and Joel Gerstl attempted to reach some generalities about engineers.[1] They were partly successful. Two of the factors that they said shape any occupation are the interests, abilities, and social characteristics of the people who enter the occupation, and the education and training that provide knowledge and values. They found, first of all, that for engineering students in general, engineering represented an advance over the social position of their parents. Engineers came from smaller towns and were of very high ability as indicated by criteria such as high school grades and general aptitude examinations, compared with other college students. The engineering students tended to spend a large amount of time in studies related directly to their major. This was not only because of the heavy requirements placed on them, but also to some extent because of interest. The engineering students in their study committed themselves to their field early and tended not to change direction once they began. They saw other courses and activities as less central to their lives and therefore were not highly motivated in other academic fields and in extracurricular activities. Compared with other students, they were highly motivated by their interest in technology and less interested in people.

Such comments are certainly consistent with the stereotypical engineer. But once again one cannot apply stereotypes to individuals. Certainly at Stanford we have a large number of engineering students from cities, and many of them come with relatively sophisticated educational and social backgrounds. We also have engineering students who are highly interested in the humanities and social sciences and others who are oriented toward people and do not like technical work. The population of engineers is so diverse that no conclusion can be applied to all individuals.

A very general breakdown of the type of work U.S. engineers do is shown in the following table. It can be somewhat misleading without an agreement on the meaning of the terms, but it at least gives an indication of the spread of the activities involved. "Development" includes the design, testing, and prototyping associated with the definition of new products. Obviously those in general management deal with people and with such functions as finance and business strategy. Those in teaching and in research live with mathematical theory and science. Those in production are familiar with solder, pneumatic wrenches, and inventory control systems. The "other" category includes the approximately 10 percent of all engineers who no longer have anything to do with engineering.

Two thirds of all engineers are in private industry, 14 percent work in education, 8 percent for the federal government, 4 percent for nonprofit organizations, 1 percent are in the military, and 5 percent work for state and local governments. Of these people, only 5 percent are female, although the proportion is higher among younger engineers, and the percentage presently in engineering school is appreciably higher. Approximately 60 percent of them have a B.S. degree, 23 percent an M.S. degree, and 5 percent a Ph.D. degree. Although college degrees are becoming the rule for engineers, many people in industry who are classified as engineers have learned their skills on the job without a degree.

Some engineers work on work stations in cubicles in huge rooms. Others work alone hundreds of miles from the nearest people. Some of them design new products. Others are concerned with getting these products manufactured, tested, and delivered. Still others sell and maintain them. Others are engaged in attempting to generate new knowledge and concepts necessary to engineers. Many of them teach. Yet engineers also worry about facilities, money, and, most of all, other people. Few engineers work alone. The majority of them either work with others or have specific responsibility for them.

The engineer is dependent on other people for knowledge, assistance, and feedback. The complexity of engineering makes it impossible for any one person to be in control of all necessary information. Engineers learn through experience, and older and wiser heads are always beneficial, if not necessary. Finally, when solving complicated problems, people can lose their perspective and become blind to errors. In engineering, a large amount of emphasis is placed on independent evaluation. These all imply interactions with other people.

Engineers in organizations require the support of many other departments and activities, such as purchasing, laboratories and shops, libraries and central computer facilities, personnel, and certainly marketing and manufacturing. Here again, interactions with other people are paramount. I began my career designing mechanical devices—machines, if you would. This activity re-

Employment of engineers by function

Development	30%
General management	19
Production, inspection	17
Management of R&D	9
Reporting, statistical work, and computing	4
Applied research	4
Teaching	2
Basic research	0.5
Other	15

Source: *Science and Engineering Personnel: A National Overview,* National Science Foundation Report NSF 85-302-5

quires a good bit of prototyping and therefore the ability to get parts made in shops. I quickly learned the procedure whereby one submitted a drawing and the proper account number to the shop, but nevertheless my parts would take weeks to appear. It took me an embarrassingly long time to realize that my jobs were generally being interrupted by work for my older and more savvy colleagues. Worse yet, the machinists were not telling me when I had obviously done something stupid. I had overlooked the fact that the people in the shop would always give their buddy's work preferential treatment. I felt a bit foolish when I figured this out, since I myself had worked in a shop and had given preferential treatment to my buddies. After I made an effort to become popular in the shop, my output took a great leap forward. We will discuss the role of engineers in managing people in Chapter 9.

Engineering also encompasses a wide range of knowledge, technique, intuition, and judgment. Some engineers are difficult to distinguish from mathematicians and scientists. They work with esoteric and sophisticated theory and powerful computers. Others are very close to technicians. They have a pragmatic hands-on approach and an almost mystical sense of rightness, working by feeling and instinct based upon experience. We can therefore count on a wide variety of cognitive styles, as well as personality characteristics. Engineering involves science, mathematics, economics, and physical resources. However, it also involves values, feelings, instinct, and luck.

Seldom does one encounter an unbiased view of engineering, because it is often described in order to suit the needs of the describer. In the press, for instance, those aspects of engineering that are most newsworthy are emphasized. We therefore hear more about computers and weapons than we do about the folks who bring us plumbing fixtures and nails. Many aspects of engineering are ignored by the media, such as the degree of uncertainty in the process, the dynamics of problem definition and public support of technological enterprises, and the interaction between science and engineering. Note that the dictionary definition above refers to the "science by which . . ." As we shall see later, engineering does not fit the common stereotype of science (actually, science doesn't either).

Many schemes are used to attempt to subdivide engineering into more understandable components. A few of them follow on pages 36–38.

Breakdown of the United States Gross National Product for 1987

Industry	Billions of dollars
Mining	85.4
Construction	218.5
Manufacturing	853.6
Durable goods	480.0
Lumber and wood products	27.7
Furniture and fixtures	15.0
Stone, clay, and glass products	27.5
Primary metal industries	36.4
Fabricated metal products	60.3
Machinery, except electrical	81.2
Electric and electronic equipment	85.0
Motor vehicles and equipment	49.9
Other transportation equipment	56.0
Instruments and related products	27.0
Nondurable goods	373.6
Textiles and apparel	42.5
Paper and allied products	39.5
Chemicals and allied products	77.1
Petroleum and coal products	33.6
Rubber and misc. plastic products	30.0
Other (food, tobacco, printing)	151.0
Transportation and public utilities	408.2
Transportation	150.8
Communications	121.0
Electric, gas, and sanitary services	136.4
Agriculture, forestry, and fisheries	94.9
Wholesale trade	313.0
Retail trade	427.4
Finance, insurance, and real estate	775.4
Services	793.5
Total GNP	4,526.7

Source: *Statistical Abstract of the United States*, 1990, p. 426

Aerospace
Agricultural
Chemical
Civil
Computer
Electrical
Electronic
Industrial
Marine and naval
Mechanical
Metallurgical and materials
Mining
Nuclear
Petroleum

Source: *Standard Occupational Classification Manual,* U.S. Department of Commerce, 1980

(1) Engineering is sometimes defined according to "industry" (electronic engineering, automotive engineering, aerospace engineering, "high-tech" engineering). Governments and business magazines like this format. The table on page 35 shows a breakdown of the U.S. gross national product by industry and the relative magnitudes of the business done by the various industries. (Notice that, although we constantly hear the words "high technology," the construction industry is over twice as big as the electric and electronic equipment industry.) To define engineering in this way is to imply that engineering activities in some industries (aerospace) are different from those in others (agriculture). To some extent this is true. Certainly more aerospace engineers than agricultural engineers use exotic computer-based structural analysis programs. However, there are also many similarities between engineering activities in different industries. The breakdown by industry is consistent with the thinking pattern of many people who collect and massage statistics, and it is loved by economists. When taken alone, however, it does not tell us a lot about engineering.

(2) Engineering is often subdivided by "field" (see adjacent list). This very popular approach allows us to talk about civil engineering, for instance, as being differentiated from chemical, mechanical, electrical, or industrial engineering. This categorization is the one that engineering schools, professional societies, and engineers themselves use to discriminate roughly among themselves. If someone asks me what kind of engineer I am, I will reply "mechanical." If the person asking the question is an engineer, this tells them something. If they are not, it tells them very little. But it seems to be socially effective.

(3) Engineering is sometimes defined in terms of the intellectual "discipline" used. Examples are heat transfer, fluid mechanics, structural analysis, control system theory, circuit design, or catalysis. Each industry in category 1 above is based on a number of the fields of engineering in category 2. Each field, in turn, is based on a number of disciplines. Mechanical engineering, for instance, includes heat transfer, fluid dynamics, engineering mechanics, and design. These disciplines themselves have many subdisciplines (kinematics, control theory, machine design, gas dynamics, turbulent heat transfer, finite element analysis). These disciplines and subdisciplines form the basis for

many books, courses, curricula, technical journals, and professional affiliations. One finds conferences and publications having to do with heat transfer, for instance, that allow people in various fields (mechanical, chemical, aerospace engineering) who are concerned with common problems to compare notes, swap techniques, and learn about new approaches.

(4) Engineering is often broken down according to activities found in technological enterprises. Here one finds functions such as research and development, design, manufacturing, management, maintenance, even marketing. More will be said about this breakdown later in this chapter and in the remainder of the book. Although simple, it is a useful one because it gives us a good indication of the variation in the work of the engineer.

(5) Engineering is also sometimes associated with a particular kind of product (integrated circuit engineering, spacecraft engineering, laser engineering, or lawnmower engineering). It is popular in the press and in common usage because it describes engineering by aligning it with a product that people are familiar with. If I tell people that I am a refrigeration engineer, they are comfortable because they know about refrigerators and what they do. However, they do not know whether I decrease the temperature of ice cubes, buildings, or rocket fuel, and they have not the foggiest notion of what I actually do in my job. It is true that we would expect an integrated circuit engineer to know more about the mechanical properties of silicon and processes having to do with deposition and photo-etching than a lawnmower engineer. But we might be surprised to discover that the two of them have a good bit of intellectual commonality in their work, despite the difference in product line.

(6) Engineering is also sometimes described in terms of industrial process. We might hear of plating engineering, ultrasound engineering, nondestructive test engineering, refinery engineering, diffusion engineering, or just plain process engineering. This breakdown is a bit more sophisticated than the one based on product, but we have to understand the process and its environment to get an indication of what the engineer does. One organization from which I used to draw my pay referred to engineers who did not have degrees from four-year colleges as test engineers. They did not especially spend their time testing things, nor were they themselves being tested. The organization

seemed to be satisfied with the title, but it had simply no value at all in communicating the nature of the work that the individuals did. They were indistinguishable from other engineers, except for their lack of educational credentials, and eventually the company would weary of the game and reclassify the individuals as engineers.

(7) Engineers are also sometimes described by their responsibilities or the entity they serve. Examples are plant engineers (responsible for the physical plant), city engineers (reporting to the city), military engineers (an arm of the military), or air pollution control district engineers. Such description tells us about their responsibilities, but once again does not enlighten us much about their typical daily activities.

(8) Finally, one sometimes finds definitions based on the intellectual problem-solving styles used. Examples might be experimental engineering, engineering analysis, creative engineering, or forensic engineering. Although such an approach is not widely used, I will occasionally rely on it in this book, because I think it helps us tune in to the types of music that play in the minds of engineers engaged in different phases of the engineering process.

Each of these approaches to categorizing engineering has value, and taken together they tell us quite a bit about the engineering process. I criticize them primarily to point out that it is not possible to describe engineering by any one simple scheme. It is a complex field and requires a multidimensional map for understanding. When I try to talk about engineering, I usually draw sketches similar to the ones on page 39. Such figures were quite common in the 1950s, when many people were worrying about what were then called the anatomy and morphology of engineering. These figures are an obvious simplification, but they are at least two-dimensional and convenient because they fit on a cocktail napkin and can be thought about without causing permanent neuronal damage. More important, they contain a number of activities that are essential to engineering. They are explained further in the remaining chapters of this book.

Engineering is interlocked with science, mathematics, and business in what we might call the environment. This environment has physical and social components. People often tend to discuss engineering in isolation, forgetting that the field can not exist by itself. Engineering itself is broken down into

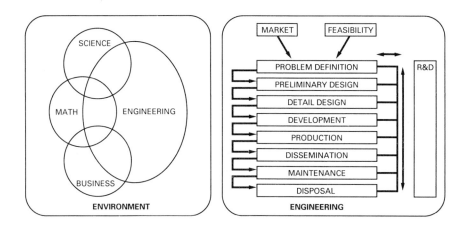

various functions, the interactions of which are extremely complex. As many people have argued, they do not simply proceed in order.[2] Engineering is an iterative process, in which it is not unusual to return to an earlier stage (or at least wish one could) and in which many stages proceed in parallel.

Although much of the remainder of the book will be concerned with better explaining this diagram, let us briefly discuss each component here, leaving engineering until last. We will then discuss a particular case—the development of computerized tomography or CT scanning—to see how the activities in the diagram interact.

Science

Science and engineering are closely related but distinctly different. As we discussed in the last chapter, through most of history science and technology have proceeded as unrelated activities. When they first became related, the science followed the technology (Toricelli was inspired by mine pumps, and thermodynamics followed the steam engine). The notion that science leads technology is extremely young. Because recent, highly visible scientific discoveries have been the basis for technological developments (lasers, nuclear fusion and fission, transistors, the structure and function of DNA), people now widely believe that technology is, by definition, the application of science. In fact, most technological activities are not based on recent scientific discoveries at all. "High" technology involves a higher than average application of scientific discovery. But high-tech companies live or die by their ability to design well, produce efficiently, and compete successfully in business. There is much more than science in such activities. Many of the activities of pure scientists (such as designing cosmological models and subatomic particle theories) have no forseeable technological application. Much engineering (including most of the fancy work I have done in industry) requires little explicit science.

Engineers do apply aspects of what is sometimes called "the scientific method" in their work. This phrase refers to the use of theory and experiment in the search for understanding. It includes an attempt to quantify when possible and to keep accurate records so that experiments can be reproduced and results verified. Engineers who have been formally educated are exposed to the same beginning science courses as science majors and acquire some of the same formal scientific techniques. Some people who are called engineers actually do scientific work (especially in academic settings, where rewards are easier to obtain through the practice of science than engineering).

The motivations of pure scientists are markedly different from those of practicing engineers, however. Pure scientists desire to understand phenomena. Their product is published knowledge, and their audience and judges are their colleagues. They are not necessarily concerned with the application of their knowledge, and they are rewarded primarily by intellectual achievement and the society and salary that accompany it. Pure scientists tend not to be found as often on the payrolls of profit-making companies, except in industrial laboratories. Even in these laboratories, though, their activities are apt to be somewhat more restricted than they are in an academic setting, because profit-making companies do, after all, want knowledge that has something to do with their business.

By contrast, engineers are motivated to solve their problem successfully within a given schedule and budget. They would prefer to understand the microscopic phenomena that cause macroscopic behavior, but they must complete their work whether they do or not. They tend to work for companies and be rewarded by promotions, by bonuses, and by the satisfaction of a job well done. Most engineers publish very little in the open literature, and in the case of the private sector and the defense industry may be discouraged from doing so for reasons of competitive advantage or national security. On the continuum between pure scientists and engineers, one finds "applied" scientists (a term that covers most scientists working in industry) and "research" engineers (many of whom have received their training in the sciences).

These disavowals of the centrality of science in technology should not be taken to mean that science is not extremely important to engineering. I make them only because of the common tendency to think of engineering as some kind of subset of science. Ideally, it would not be necessary to separate the two. If we insist on doing so, however, we should be consistent enough to give each its own identity. Certainly science is invaluable to modern engineering. Not only do engineers use many of the same intellectual approaches as scientists, but a very large number of theoretical concepts in engineering have been taken from science, and scientific understanding of phenomena is of great value to engineering. Engineering, of course, also strongly affects science. No branch of science is without sophisticated observation and measuring equipment and computer reduction of experimental data. In addition, engineering activities often illuminate areas of ignorance and uncover interesting behaviors. The development of solid-state electronics is an excellent example where scientific knowledge is essential to the engineering and where engineering activities tend to focus and direct scientific investigations.

Mathematics

Mathematics is the theoretical language of technology. It is perhaps the largest barrier to understanding science and technology. Although everyone has been exposed to mathematics, few people have a sophisticated understanding of it, and many people, if given a chance, will avoid it entirely. Mathematics is important in engineering not only because engineers must think in terms of quantity but also because it allows the modeling of physical relationships that in turn allow analysis, optimization, and prediction. Math is the intellectual

$$\text{div } \mathbf{E} = \frac{1}{\epsilon_0} \rho$$

$$\text{div } \mathbf{B} = 0$$

$$\text{curl } \mathbf{H} = \frac{\partial \mathbf{D}}{\partial t} + \mathbf{J}$$

toolkit that separates the engineer from the technician. It is essential for the degree of sophistication necessary to design airplanes, high-speed printers, and oil refineries. It is less necessary in the design of automobile jacks and hair dryers.

As with scientists, there is a wide gulf between so-called pure mathematicians and engineers. Mathematicians seek to find intellectual order. Their concerns are with new insights. They are motivated and guided by aesthetic considerations such as elegance, rigor, and beauty, and their forum is their peer group. A large proportion of mathematics has found powerful applications, but this was not necessarily among the considerations that motivated the work initially.

A middle ground between mathematicians and engineers, analogous to that in science, is among applied mathematicians, who may have a background in mathematics, in a field of engineering, or even in science. More theoretical engineers may have high mathematical sophistication. However, most engineers use mathematics for what it can do. They may appreciate it, and even like to work with it, but its prime value is the power it gives to the engineer, not its inherent beauty. I learned this painful distinction as a graduate student. I had taken many mathematics courses at Caltech and Stanford and had always received A's and had been at the top of the pack in exams, so I considered myself a reasonable mathematician. But the last mathematics course I took in school was on the topic of variational calculus, and I enrolled in it because it was taught by a very good mathematician who was considered a legendary teacher by my mathematician friends. His lectures were most impressive, but on the problems I kept getting very low grades, even though I had gotten the "right" answers. Along with my very low grades came a continuous stream of comments such as "inelegant," "nonrigorous," and "ugly." I proved I was an engineer by dropping the course, rather than learning to produce the requisite beauty.

Mathematics is essential in modern engineering. It is, of course, also basic to the physical sciences. In physics, theory is often based on mathematical relationships. This is particularly true in areas such as cosmology, which is concerned with the nature and origin of the universe. It is simply not possible to conduct experiments that reproduce the birth of the universe or simulate

black hole. Mathematical models that are consistent with
sical phenomena are the basis for cosmologists' beliefs
f the universe. Mathematics is also becoming more central
d life sciences. It is also finding increased applications in
business.

Business

In almost all technological situations, it is necessary to hire, train, and manage people and to control work and resources. Private companies have to consider profit, accountability to stockholders, and investment. Organizations supported by public funding are accountable to the public. Since technology usually requires significant resource support and involves large numbers of people, business considerations are integral. Functions such as marketing, management, and finance cannot be separated from engineering in a healthy technology-based company. In organizations whose output has to do with sophisticated technology, it is not unusual to find people in general management positions who have engineering backgrounds. Many young engineers believe they have an aversion to business, but those who maintain this aversion eventually find themselves restricted in what they can accomplish in their careers.

The Environment

What is the environment in which engineering operates? Thoughts often turn first to the natural, or physical, environment. Over the past forty years, and especially over the past twenty, we have become very aware of the relative fragility of what we call nature. Through technology we have modified nature to better serve our materialistic desires, but in so doing we have destroyed enough of it to now begin to worry about not only aesthetic factors but dangers to our health and longevity. Engineers are now being expected not only to find ways of repairing past damage but also to operate in such a way as to drastically decrease future damage. (I recently saw a newspaper article featuring ten suggestions from various engineers on combatting the greenhouse effect, which is the warming of the atmosphere owing to the increasing amount of carbon dioxide. One of them that particularly caught my eye involved maintaining a giant fleet of aircraft in the air to spread materials that would decrease the intensity of solar radiation enough to offset the warming of the greenhouse effect. Somehow this technological solution seemed flawed.)

The real environment in which engineering must operate is not the physical environment but the human environment. It is the desires of *people* that lead to products and services that despoil the physical environment. It is also the desires of people that now cause us to worry more about it. People are the purchasers of the outputs of technology, and that purchasing power casts the final vote on the directions technology takes. Through increasing attempts to regulate technology (discussed in Chapter 10), people are demonstrating their concern that technology be consistent with social and cultural values.

Engineering

The engineering process begins with a desire. This is reduced to a problem. The process by which problems become defined and attacked by engineering is important and somewhat mysterious and is discussed in the next chapter. The process can only be successful if reasonable knowledge is available (it is not much fun to work on antigravity these days) and if someone wants the result. In the diagram these are referred to as technical feasibility and market.

After the problem is defined, an activity sometimes called "preliminary design" is undertaken. Various ways of solving the problem are conjured up and compared. Toward the end of the process, a decision is made as to which design to pursue, and the overall configuration of the product is defined. This is often (and erroneously) considered to be the creative phase of engineering (erroneous because creativity is required in all phases of engineering—more about that later).

Preliminary design evolves into detailed design, in which the product is completely defined in its final form. This means that each component of the product is described so that it can be produced and so that the whole will fit together into a balanced and integrated product. This process requires an understanding not only of the product but also of materials and the way that they can be formed and of assembly techniques. At one time in history, detailed design was often done by trial and error, in shops and labs. Now as much design as possible is done on paper or in the computer. Each part of a product is completely specified as to material, shape, size, and pertinent manufacturing and assembly processes.

No matter how intelligent the people are, when a product is first produced it usually does not work the way people expected. It may have flaws, or "bugs," its performance may be less than desired (or more), it probably costs too much to produce, and it probably fails in unexpected ways. In an activity sometimes called "development," prototype products are brought to the functional and economic level that will satisfy the customers, whoever they may be, and the producers. This process is well known to engineers, and less so to those outside of the field. It is not highly publicized, since neither engineers nor customers are happy with the fact that people seem to make mistakes, things seem to fail, and knowledge is not perfect. The process of development makes extensive use of test, measurement, and other iterative approaches.

The process of development, like design, varies widely among different engineering fields. For instance, in activities where the scientific base is high, such as microelectronics, theory may be very helpful in spotting the sources of problems and defining changes. In other areas, such as mechanical design, development is often necessary because the theory is inadequate. For instance, no physical law allows us to predict how long things will last before they wear out. It is necessary to build prototypes and test them to see. This phase of engineering is important to emphasize, because more and more people seem to be developing unrealistic expectations of engineering. They want to believe that it is possible to build extraordinarily sophisticated new technological products within a predictable schedule and budget and have them operate perfectly from inception. The Strategic Defense Initiative (Star Wars) is a recent example. Despite the many statements of experienced people that it is impossible to build such a system and achieve the desired reliability without full-scale testing, the government and many sectors of the public and industry retain their optimism. Unfortunately, my personal experience does not allow me to be one of them.

The next function in engineering is production. This is the process by which the debugged design is made for the customer. There may be only one product (the Golden Gate Bridge), a few (space shuttles), a large number (Cadillacs), or a bewildering quantity (rubber bands). If a product is to be manufactured well, production considerations must enter into the preliminary and detailed design stage. This was not obvious to American industrialists for a long time. Today the United States is being forced by foreign competition to devote a major effort to assuring that production considerations are more effectively integrated into design. One of the problems has been that engineering schools are relatively weak on production, but that situation is beginning to change.

After production, the output must somehow be disseminated to the users. In the case of the Golden Gate Bridge, or even the space shuttle, this is easy, requiring only transportation and people. In the case of Cadillacs or rubber bands, a network of outlets and logistics and marketing efforts is required. Even after delivery of a product into the hands of the customer, maintenance and service are being demanded as never before. This consumer demand requires not only more consideration of reliability and serviceability during the design and production of the product but also the provision of maintenance capability in a form that is acceptible to the customer.

Finally, for environmental and economic reasons, disposal of the products and by-products of engineering is becoming more of an issue. We are no longer oblivious to landscapes littered with nonrusting beer cans, to water sources contaminated by industrial chemicals, or to nonbiodegradable plastics that release carcinogens when burned. More and more, engineering must consider the final destination of its products.

The last topic in this brief overview of engineering is research and development—the investigation carried out specifically to produce new knowledge and products. This is an essential function in engineering and one which overlaps other activities to a great extent. It is receiving a great amount of attention these days because the United States' lead in research and development is eroding. We still invest by far more money than any other nation, but our per-capita figure is slipping. In addition, as will be discussed later, the majority of research and development work in this country is military, as opposed to that in countries such as Japan and Germany. Military research, we are learning, does not have a large enough economic impact on international competitiveness. Greater ability to compete economically requires that we focus more of our engineering research specifically toward that goal.

This breakdown of the engineering process is a simplification, but it will allow us to discuss engineering in a somewhat orderly way. In actuality, the various portions of the diagrams on page 39 are parts of a continuum rather than separate activities, and it is easy to find examples where engineering projects violate the diagram. One can find products that make no sense from a business viewpoint, have nothing to do with science, or violate the environment. It is possible to find products that were first built in shops and then put on paper. It is possible to find examples of products that were completely designed, only to be discarded and the process begun again. It is almost always the case that the various phases of the engineering process overlap and that each contains aspects of the others. However, this scheme is a place to start.

An Example—CT Scanning

Let us briefly discuss the development of CT scanning (or, formerly, CAT scanning), now a routine, powerful, and expensive medical diagnostic procedure, as an example of a few of the interactions we've just outlined.[3] CT stands for computerized (or computer-aided) tomography. It exposes the patient to a narrow X-ray beam in a very large number of directions. The data from these exposures, rather than going to a film, go to a computer, which then generates images of the patient. The advantages are many. The computer is able to produce cross-sections of the patient at any desired point in any desired plane, and the resolution obtainable is extraordinarily high compared with that obtainable with a single X-ray source and film, since the computer can enhance a given feature, vary "contrast," and otherwise manipulate the image. The following illustrations show a CT scan cross-section of the head and the abdomen. Notice the optic nerve, the gray and white matter definition

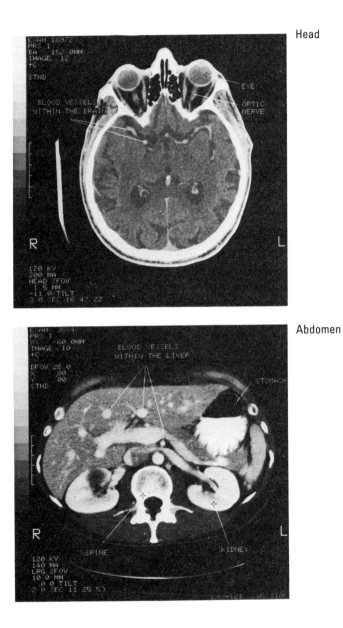

Head

Abdomen

CT scans of head and abdomen. These images were produced by the General Electric CT9800 Quick System.

in the brain, and the detail of the spine structure. CT also allows the physician to see minute variations in tissue density that can indicate early tumors and other unwanted changes.

The CT scanner depends on X-rays, which were discovered by Wilhelm Roentgen in 1895 while he was studying the motion of what were then called cathode rays (now electrons). Actually, the physicist J. J. Thomson had noticed that glass tubing held at some distance from a cathode ray tube flouresced just as the tube itself did, but he did not seek to discover why. Roentgen covered a cathode-ray tube with black paper in order to attempt to isolate the reason for reaction, and he discovered that the glass tubing would continue to flouresce. He had discovered X-rays.

It was several years before scientists understood that X-rays were electromagnetic radiation, like visible light, only of a much shorter wavelength, or higher frequency. At the time of Roentgen's discovery, quantum physics had not yet become established, and scientists did not yet realize that charged particles that are accelerating or decelerating emit electromagnetic radiation. (The same process produces radio waves from a broadcasting antenna. These are also electromagnetic radiation, but of a much longer wavelength.) In the tube used by Roentgen, electrons were accelerated by approximately 10,000 volts; when they suddenly decelerated by hitting a copper target, X-rays were produced. The X-ray machines used by dentists and doctors today still use a similar tube. The radiation from these tubes is allowed to escape through narrow slits in a shield of lead or other heavy metal and pass through your teeth, body, or whatever. They interact more strongly with heavy atoms (calcium, as in bones or teeth) than with lighter ones (such as those in tissue), and since X-rays will expose photographic film on the other side of the target, they produce an image whose density is proportional to the weight of the atoms they have passed through. In the case of CT scanning, therefore, science developed one of the precursors of a technological invention by discovering X-rays. Technology then reduced this discovery to practice, in the form of the X-ray machine, and the medical profession became dependent on it as a diagnostic tool.

Mathematics was also essential to the CT scanner and central to its invention. A problem pertinent to X-ray diagnosis might be the following: If we know the X-ray attenuation coefficient at each point within the body, could the total attenuation of an X-ray beam be calculated for any path through the body? The answer is yes. But mathematicians are fond of what are called inverse problems, and so they went on to ask this question: If the attenuation of an X-ray beam were known for a number of paths through the body, could the distribution of attenuation coefficients be calculated? The answer again is yes. This problem was solved by a Viennese mathematician named Radon in 1917. At the time, no one made the mental jump to thinking of attempting to produce an image of the cross-section of a body by exposing it to narrow X-ray beams from a large number of directions, measuring the attenuation of each beam, and using Radon's formula to calculate the precise attenuation coefficient at each point inside the body. These coefficients would have given an important image for diagnosis, since not only do bone and tissue have very different attenuation coefficients but so do such things as tumors and blood clots. Even better, since conventional X-ray images are limited because they compress a three-dimensional image onto two dimensions, this approach would allow the construction of a two-dimensional image of a three-dimensional object (a cross-section of the body) with much improved discrimination. However, even if someone had thought of this, it would have been impractical to implement, since computers did not exist and the calculation of the quantities of interest and presentation of the results are too complex to be done by hand. Even though the science and the mathematics existed, the technology did not.

Forty years later, someone finally made the mental leap; in fact several people made it. One of them was a man named Allan Cormack, a physicist educated at Cambridge University and on the faculty of the University of Capetown in 1956. During that year, the Groote Shuur Hospital lost its regular physicist and asked the university for help. As a result, Cormack spent 1½ days a week at the hospital doing the work of the radiological physicist. This experience impressed him with the fact that more accurate values of X-ray attenuation were needed. He wondered whether it would not be possible to do better than the classical X-ray by using a number of views taken at different angles to reconstruct the attenuations within the body. He was not familiar with the work of Radon.

At the end of the year, Cormack went on leave to Harvard University and developed his own mathematical theory for image reconstruction. Upon his return to South Africa in 1957, he tested his theory with a laboratory simulation. He built a circularly symmetrical test object, of aluminum and wood, and used a collimated beam of gamma rays and a Geiger counter for a detector. He moved the object through the beam in 5 mm steps, measured the attenuation each time, and then mathematically processed the data to obtain the attenuation coefficients across the test body. His results correlated nicely. This was classic applied scientific research.

Later that year Cormack moved to the faculty of Tufts University in Medford, Massachusetts, and continued his experiments. He developed an improved mathematical approach, tested his scheme with more complex models, and used computers for reducing his data. Although he showed his results to a number of radiologists and published his findings in journals, he was not able to find any interest in his work.

In 1967 a man named Godfrey Hounsfield began similar work at EMI Ltd in England. He was an engineer working on pattern recognition and was not familiar with either Radon's or Cormack's work. However, he too realized that mathematical techniques could be used to reconstruct the internal structure of a body from a number of linear X-ray measurements. He also created a useable mathematical relationship and theoretically concluded that such an approach should give the local X-ray attenuation coefficients inside the body to an accuracy 100 times better than traditional X-ray techniques.

Hounsfield had the advantage of backing by a large corporation. His first experiments were similar to Cormack's but soon he began using an X-ray tube and nonsymmetrical test bodies. He reduced the imaging time from 9 days to 9 hours, succeeded in winning the interest of the British Department of Health and the cooperation of two local radiologists, and was soon scanning sections of the brains and carcasses of animals.

Cormack and Hounsfield won the 1979 Nobel Prize in medicine for their work. They were unusual recipients of this award, since neither of them had backgrounds in biology or medicine and their work was in applied research rather than the usual basic life sciences. But the CT scanner had an extraordinary impact on medical care, and their independent efforts branded them as the originators of the concept. Their work also is a good example of successful applied research and of the ability of science and mathematics to contribute to the invention of a new technology. What else does the CT scan story tell us about the figures on page 39?

The environment was very favorable for the development of CT scanning in the late 1960s. The more economically advanced nations were becoming committed to improving health and medical care. People were demanding the best, and in the wealthier nations governments and private agencies were responding with benefit and insurance programs that made high-cost diagnostic procedures economically feasible for a wide range of people. The rapid development of the computer and electronics along with the increased sophistication of medical training and medical practice were making great strides possible in the application of technology to medical care. In 1967 the problems of skyrocketing medical costs and allocation of limited health-care resources had not yet been acknowledged. It was a time of great optimism and great capability.

On the business side, medical engineering was a growth area. EMI was a large company that had emerged from World War II with great strength in electronics. Yet its major commercial successes had been in music and entertainment (the automatic record changer, stereophonic records, magnetic recording tape, the acquisition of Capitol records and the Beatles). By the time the company was attempting to better use its capability in electronics, Hounsfield's research and development work had reached the clinical stage. A machine for head scanning was installed at the Atkinson Morley's Hospital in Wimbledon in 1971. It moved a single source and a single detector 1 degree at a time until 180 projections were taken. It took 4.5 minutes to gather the data, followed by 20 minutes for construction of the image.

Much discussion ensued among EMI's top management about what to do with the scanner. To pursue it would give EMI a new product direction. But the company had no experience in medical electronics, and the largest market for CT scanners was projected to be in the United States, which was outside of EMI's usual electronic marketing territory. However, changes in top management resulted in an increased desire for diversification at EMI and greater attention to a few promising R&D projects, of which CT scanning was one.

Early EMI estimates concluded that the worldwide market would not exceed 25 scanners. Yet by 1972 one leading American neurologist was forecasting that at least 170 machines would be required by major U.S. hospitals alone. Since the price of EMI scanners at this time was projected to be about $400,000, this was clearly a sizeable business, and EMI decided to invest in the endeavor and to charge ahead. The company was confident that the work they had done, coupled with patent protections, would give them three or four years in which to establish a solid marketing organization and gain a major share of the market.

The technical feasibility and market seemed to exist. EMI's engineering department had undertaken a large number of prototyping and feasibility investigations during the research and development phase. It was now necessary only to design a commercial unit and produce it. The decision was made to rely heavily on outside sources for the technological expertise that EMI did not have and to focus on the integration of the system at EMI. Outside vendors would account for 70–80 percent of the scanner's manufacturing. EMI's scanner reached the market in 1974, and by 1977 had brought EMI pretax profits of 12.5 million pounds on sales of 42 million pounds and had resulted in a backorder of more than 300 units.

Somewhat to EMI's surprise, however, sizeable and rapidly moving competition entered the scene. By the end of 1974 two other scanners were announced, one by Digital Information Sciences Corporation and one by Ohio Nuclear, which achieved a more detailed image than EMIs, used a 2.5 minute scan rather than a 4.5 minute one, and was priced below the EMI scanner. EMI responded by announcing a second-generation machine that used multiple beams of radiation and multiple detectors, allowing 10-degree rotations rather than 1-degree and cutting the total scan time from 4.5 minutes to 20 seconds. In addition the image resolution was increased by a factor of four. The machine was unveiled in May 1975. However, in 1975 Pfizer, a large European company, bought the rights to manufacture the Digital Information Sciences machine and six new competitors (Syntex, Artronix, Neuroscan, General Electric, Picker, and Varian) entered the scanner market. Of particular note was a third-generation machine shown at the annual Radiological Society of North America (RSNA) meeting in December 1975 by General Electric. This

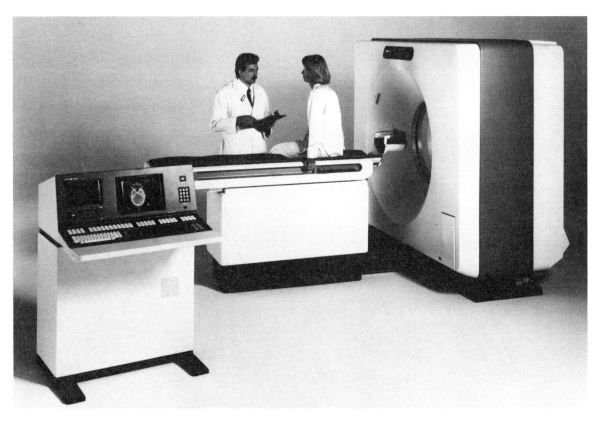

The General Electric CT 9800 Quick System.

machine used a 30-degree wide X-ray beam, which completed a single continuous 360-degree scan in 4.8 seconds. Such rapid scanning improved the image by minimizing blurring caused by movement of the patient. General Electric had engineered themselves into a technological lead. Worse yet for EMI, GE also had a U.S. sales force of 300 people and a service network of 1,200 and was intending to spend $5 million annually to retain its technological advantage.

The companies in the CT scanning business were obviously competing on the basis of technological sophistication as well as price and reliability. Complex mathematics and computer application and continued scientific research were necessary to achieve this sophistication. The environment valued performance, and the level of competition was causing the technology to advance at an extremely rapid rate. The customers seemed to be willing to pay the price (GE's original machine was priced at $615,000) if they could receive the performance. The business was also growing. It was estimated that some 150 new scanners were installed in the United States in 1975 alone.

What was happening to EMI? Through 1976 EMI had sold 450 of the 650 scanners purchased in the world, but during 1975–76 their market share had dropped to under 60 percent. In December 1976, at the annual RSNA meeting, sixteen new competitors exhibited scanners. The issue of escalating health-care costs was also being discussed, and more and more agencies were putting restrictions on governmental and private expenses. The size of the market was also resulting in tensions within EMI as well as production problems. By 1979 EMI was losing money on scanners, and it finally sold its scanner business to General Electric for $37.5 million.

As far as EMI was concerned, this was a story of an organization inventing a product, developing the original technology, becoming dominant in the business, and then losing out to a company with more resources as the environment changed. It is an excellent example of the interaction of engineering and business. EMI did the original and essential engineering. Early competing scanners were extremely similar to the EMI machine, since the patent protection was not as strong as EMI had originally thought. But the original engineering was not good enough to withstand the competition, and the final result was that EMI went out of the business.

The General Electric CT 9800 Quick CT system (see figure, page 54) is capable of performing a normal CT scan every 15.5 seconds, using a scan time of 2 seconds or less. Each scan results in approximately a million detector readings, which the computer translates to an image. Display features include extremely rapid pan and zoom of the image, simple contrast adjustment, the reformation of axial images into oblique or other views, and superposition and contrast of various images by means of a split screen. Options available with the system include three-dimensional imaging (see figure, page 55), blood-flow imaging, and bone-mineral-density imaging.

Today, magnetic resonance imaging (MRI) has taken over some diagnostic tasks from CT scanning. It does not subject the patient to X-radiation and is better able to discriminate certain details of body structure. It is also sensitive to quantities other than density, such as phosphorus concentration. Like CT

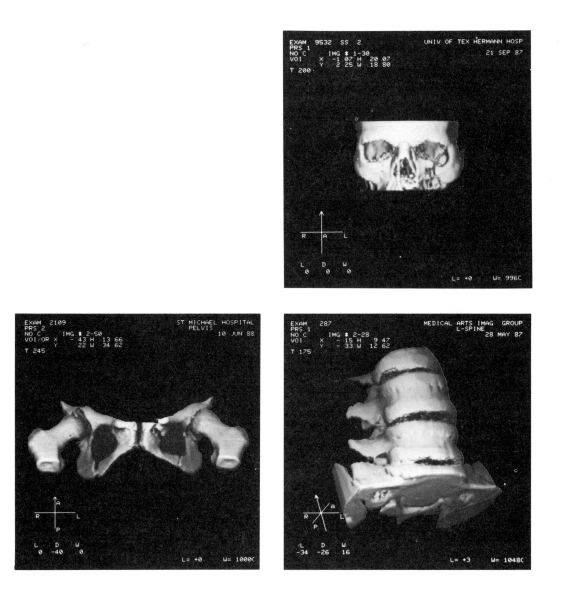

Three-dimensional images of the skull, pelvis, and spine.

scanning, MRI depends on images constructed by computer calculation. The information used by the computer depends on a physical phenomenon called nuclear spin resonance. Atoms that contain an odd number of protons and/or an odd number of neutrons act as tiny magnets, due to the fact that their nucleus contains a net charge spinning around an axis. If they are placed in a magnetic field, the spin axes of these atoms will align themselves with the field. If they are then excited by a burst of electromagnetic waves of the proper frequency (the resonance frequency), they will begin precessing, or wobbling about their axis, with a frequency that is proportional to the strength of the magnetic field. In doing so they will emit electromagnetic radiation at this precessional frequency. In MRI, the patient is placed in a very strong magnetic field that varies in strength over space. The proper signal then causes the now-aligned hydrogen (or whatever) atoms in the patient's body to precess with a frequency that is a function of their location. The intensity of the signal translated by the precessing atoms at any location and the rates at which this signal decays back to its pre-excitation state can be used by the computer to determine the concentrations of the atoms of interest at that point and other interesting quantities.

Many other types of imaging used in medicine depend on sophisticated technology. Some of them, such as those used to check on fetuses in utero, use acoustic waves (ultrasound). Quite a bit of attention is presently being paid to Positron Emission Tomography, or PET scanning. In this procedure small doses of radioactive glucose are injected intravenously to indicate on a monitor where blood flows. Researchers are using it to learn more about the chemistry and pharmacology of the brain, since parts of the brain become engorged with blood when they are highly active. CT scanning will continue to occupy an important niche in medical diagnosis, but the extent and shape of that niche will depend on technological developments in all areas pertinent to medical diagnosis.

We will now look at some of the more significant aspects of engineering in detail. The next chapter will discuss an important stage in the engineering process that receives altogether too little attention: the origin of problems. Why do engineers work on the things they do?

3

The Origin of Problems

The Pushes and Pulls

Most people wonder, from time to time, why technology takes the directions it does. Engineers face this question often. Why is so much technology oriented toward war? Why don't you engineers invent a better way to trim hedges? prevent auto accidents? neutralize toxic wastes? Who on earth needs more computers? micro-processor-controlled toasters? electric toothbrushes? In this chapter we will take a brief look at the origins of problems for the individual engineer, and then turn to the factors that determine directions for engineers in organizations. Finally, we will consider the directions of the organizations themselves.

Occasionally I ask engineering students to tell me what kind of job they think will give them the most freedom to do the sort of engineering they really want to do. Their answers are most often "start my own company," "work for a very small company," "become a consultant," "become an inventor." In other words, self-employment or employment in a small company is seen as allowing more freedom than working in a large organization. Yet the majority of engineering students do join large organizations upon graduation from school. What is the attraction of self-employment or work in a small company?

To the extent that economics will allow, people who work as independent consultants, inventors, or entrepreneurs have considerable choice not only in the problems they take on, but also the environment in which they work, the schedule they set, and the colleagues with whom they associate. Even these engineers are not without constraints, however. Unless one is independently wealthy, one must engineer what people want; and even if one were wealthy, I should think that it would be frustrating to produce ideas and products for which there were no customers. Independent engineers are also restricted to jobs that can be done with the limited resources available to individuals and small groups; they cannot make commercial airliners.

I was amazed, when I had my first experience with hiring engineers, to find that consultants were often quite willing to put aside their romantic-seeming life and come to work for a regular paycheck. I had always worked for companies and had occasionally fantasized about being professionally on my own. But I later learned that the life of an independent engineer is not that of Adam and Eve in the pre-apple garden. In *The Atom and the Fault* Richard Meehan, a friend of mine and a geotechnical engineer who runs his own consulting company, discusses his involvement in problems having to do with locating nuclear reactors in a state riddled with earthquake faults.[1] Besides describing the fascinating confrontation between the power companies, governments, property owners, and environmental groups that defined the nuclear industry (or lack of same) in California, he talks of such problems as gaining weight because of the interminable lunches one must spend with prospective clients, worrying about maintaining a workload that is steady enough to reliably feed an office, and dealing with his clients' desire for safe and efficient solutions rather than brilliant ones. In *Getting Sued and Other Tales of the Engineering Life,* Meehan discusses other problems, as the title implies. Working as an independent has its advantages, but all engineers must please the customer as well as play with technology, and the customers of consultants are not from a more enlightened planet.[2] I personally prefer to combine my consulting with working for a salary.

The university in which I presently work is similar to some government, company, and nonprofit research laboratories in that it offers the engineer an extraordinary amount of independence in an organizational setting. I have been a professor at Stanford for a long time and have yet to receive a direct order. University faculties simply do not operate in that way. Obviously, what I do in the classroom and in administration contributes to the success of my department, my school, and the university. Yet I have a say in defining what I do. A portion of my life (my research and consulting) is completely up to me. At the same time, I receive a paycheck and company benefits, have a large number of fascinating people to play with, and have a great amount of security. Is this the perfect home for an engineer?

Would that it were. First of all, the university environment severely restricts the type of engineering one can do. Universities do not make dams, detergent, or automobiles. They do very little design and no production. I cannot be part of a large project team, an activity which I love. Neither can I work my way up to becoming the godlike president of a large technology-based company. Second, I am expected to bring in reasonable amounts of money, in the form of grants, gifts, and contracts, to support my work. Good citizens of the Stanford Engineering School cover part of their own salary, stipends for graduate students, and the cost of necessary equipment, supplies, travel, and

administrative expenses for their research. Do they have to raise money, since they are tenured? No, but it is not pleasant to be viewed as not carrying one's weight. There is a cost to such behavior. Tenure is merely legal; if one violates the values of any society, the society will usually find a way to make one suffer. Finally, although I am free to define my research, my life as a professor is better if my activities cause people to respect me intellectually and if they contribute to my field. For what it is worth, I also find the university more intense than industry. It is true that deadlines and expectations are largely self-inflicted by faculty members, but good universities are smart enough to hire people who are compulsive achievers. I may be typical, in that I work harder if I am defining my activities than if others are. Finally, universities are very, very conservative institutions. They offer the individual great freedom, but as organizations they tend to move very slowly and follow social trends rather than lead. They are structured in such a way that new directions require a consensus of professors with individual interests; the result is tremendous inertia.

The circumstances of my life are somewhat similar to those of engineers in many research laboratories. They have freedom to define their work but are expected to do work that benefits the laboratory. Should the likelihood of such behavior not be high for a prospective employee, that engineer would not be hired in the first place. Universities and research laboratories are wonderful working environments for certain types of engineers, but even these half-way houses between independent practice and traditional companies also place constraints on one's options.

Life in a small start-up company, Cygnus Therapeutic Systems, which manufactures transdermal drug-delivery systems—adhesive patches that allow pharmaceuticals to enter the body at a steady and controlled rate. Top: Gary W. Cleary, founder, chairman of the board, and chief technical officer in his garage working on the process that was the basis for the company. Middle: The original four employees, celebrating their first paycheck. Bottom: One of the company's first production machines. Cygnus Therapeutic Systems is now a successful and rapidly growing publicly held company.

The majority of engineers work on the problems they are told to work on, in large organizations with strong management. They work for salaries and have bosses. A successful life in most organizations allows one to argue, but eventually one either goes along with one's boss, changes the boss's mind, or leaves. The formalization, discipline, and control are necessary in order to allow the organization to be stable and to allow projects to be broken down into jobs that can be accomplished within budgets and time limits. Organizations accomplish much more than mobs, and large projects offer challenges and satisfactions that smaller ones cannot. Tight control of complex assignments also provides good training for young engineers and assures that they work within their level of competence. Life in a complicated organization necessarily includes bureaucracy, however, and it occasionally leads to mediocrity, duplication of effort, and frustration for creative people.

I first got a taste of this when I worked on the design staff of a large U.S. automobile company early in my career. I was in the midst of large numbers of talented and motivated young designers who were highly creative and quite sensitive to transportation problems but who found it very difficult to affect the final products. Automobiles seemed to be formed by some sort of unyielding corporate mentality. There was an obvious market for small cars that were quick and nimble, cars tailored for commuters, and cars that would better serve people with various sizes of families. We were given enough money to experiment with such concepts, but when the actual products emerged they were the same old stuff—different flavors of soup rather than different foods to eat. In fact, the company policy at the time was to decrease the number of different body and drive-train parts they produced, narrow the product line, and create the impression of diversity through cosmetics and advertising. During this time, I had great difficulty keeping a straight face when listening to car owners debate the relative merits of the automobiles produced by this company, since under the skin they were quite similar indeed. At that time, since Detroit had no serious competition in the United States, such a move probably did maximize profits, but it was most discouraging to many of the engineers, especially the young ones.

I spent my most frustrating period in an automobile design studio that was producing an "advanced" prototype, which is the traditional way the automotive industry thinks about future products. Many of us young designers were from southern California, and even back then we were aware of problems having to do with smog and congestion. We loved cars that handled like a dream, were cleanly and beautifully designed, and got reasonable mileage. We knew that a century into the future people were not going to be moving around in two-ton metal boxes that guzzled gas and released waste products into the atmosphere. Our views were humored but finally ignored. The finished prototype was extremely heavy, horribly complicated, and voracious in its appetite for fuel, and it looked somewhat like the offspring of a mating between an airplane and a shark, but unfortunately with neither the performance of the former nor the grace of the latter. It was another in the continuous series of sleek showcars that the company produced that had little to do with actual transportation.

Large organizations do not move fast, and when they have a successful product line, as the automobile company did at the time, they are extremely hesitant to experiment. Obviously this inertia is not only sometimes frustrating to engineers but in the long run it can cause troubles for these organizations when the world changes, as has occurred in the automobile business. Japanese automobile companies design cars quickly with small teams, use standardized components to explore a wide variety of configurations, and have an approach to production which uses small inventories and extreme flexibility of manufacturing and assembly. Detroit is learning, but it is not changing its stripes easily.

Other problems can arise in large organizations because engineers may be so far removed from decision points that they tend to follow orders without questioning the source. Groups of talented people may run themselves ragged on chases that are later found to be in the wrong direction. This is a built-in flaw in well-running organizations. Engineers who question their assignments are valuable, because they do produce new approaches and ideas. However, they complicate life for managers, increase the uncertainties in projects, and introduce risk into the technological process. In a large organization that lives by producing products, one tends to follow orders.

I spend quite a bit of my time consulting for companies interested in increasing their creativity and responsiveness to change.[3] One very straightforward approach I take is to encourage employees to analyze the problems they work on. Are they the important ones? Is the problem stated in a way that implies a direction or answer, and if so is the implied answer or direction reasonable? Does the statement of the problem include other inhibiting constraints, and are those constraints valid? I try to teach engineers not to accept the problem exactly as stated to them—to back up a step and look at what is wrong or what is needed in a more general sense. This approach results in more creativity, although it also produces more ripples in the water. It is easier to do as a part-time consultant than a full-time employee, since risk is involved. As an employee in a large organization, one cannot lose by solving the problem exactly in the way assigned by one's superior. However, one may pass up opportunities for a more successful solution of the problem.

Organizations are conservative in their problem solving because mistakes cost lots of money. Much effort is therefore spent in checking and re-checking. Tried-and-true approaches are preferred over new ones. Engineering, like most activities, also entails administration and politics, and the larger the organization, the more time is consumed in dealing with such things. Engineers, especially young ones, may therefore not have much time to experiment. This is one of the major costs of working in a large organization, but it is offset by benefits: larger scale products, more varied colleagues, and nice "perks."

A common way that engineers make their work in large organizations more varied and rewarding is to move between and within organizations. Many bosses who have risen through the ranks themselves can appreciate the need of employees for variety as well as autonomy in their jobs. Organizations can also be surprisingly sensitive to the personal values of their employees, and often take those values into account when making job assignments. I have refused many assignments, and my career has not suffered; however, I have also accepted many assignments that I did not like, out of some combination of organizational idealism and loyalty. I have not worked on a project that gave me significant ethical concerns, but perhaps I have just been lucky.

The specificity of problems assigned to engineers varies tremendously. In the early stages of a career, one is generally told what to do in great detail, and one is glad of it, since school teaches very little in the way of practical engineering. ("Jim, design a stamped sheetmetal bracket to hold this stoplight to the chassis. Here are drawings of the stoplight and the chassis. Joe Eberly is an experienced designer and will help you if you have problems.") As one becomes more experienced, problem assignments give one much more latitude. ("Jim, we are putting together a preliminary design team to define a new gas chromatograph. You are the mechanical engineer on the team.")

As engineers acquire management experience, they are often put in the position of defining problems for other people. This requires the ability to diagnose situations and break a broad problem into more specific ones. As engineers work their way through the ranks, they are called upon to make product decisions and decisions having to do with the budgeting of people, time, and money. Eventually, they may become active in setting company goals and strategy and assessing performance. Do they ever get to do the engineering of their dreams? Sure, assuming they position themselves cleverly and their dreams make enough money to satisfy the financial needs of the organization.

The projects an organization chooses to pursue heavily influence the lives of engineers within the organization. Organizations have a number of somewhat conflicting desires. They would like to make as much money as they can. They would like to do things that enhance their reputation. They would like to do things that are satisfying to their employees. They would also like to work cheaply enough so that they can compete. If a company tries to increase its income from sales by constantly raising its price beyond the competition, it will die. If it tries to increase the number of units sold by lowering its price below cost, it will die. If an organization chooses work that fails to stimulate its employees or violates their values, it will fail through internal dissension, lack of motivation, and inefficiency. If it selects projects solely for their motivational and social value, it may not be able to pay its bills.

Most projects that a company chooses to take on are a compromise. They are sophisticated enough to challenge the engineer but are not at the cutting edge of technology. Occasionally, as in the early days of U.S. space exploration or in the labs of extremely successful companies, one can find the resources to explore and experiment to one's heart's content. However, more

often in a company one finds enough money to complete projects but not enough to do them as well as one might wish.

Organizations in the private sector choose projects either to make money right away or to position themselves for future profits. These projects must obviously be consistent with the available resources, be supported by the necessary knowledge base, and be oriented toward a reasonable number of customers. In organizations that are nonprofit (research laboratories, universities), projects are often outlined and then proposed to a funding agency, either private or public. Negotiations then result that determine whether the project will be funded and at what level. Government organizations usually select projects that correspond to their mission (develop a new pesticide, do research on low-speed aerodynamics, or develop a more powerful free-electron laser).

The work that engineers do is affected by the type of funding. For instance, companies in the private sector have difficulty justifying research that cannot be connected to profit in the reasonable future. The government, on the other hand (through agencies such as the National Science Foundation, the Department of Defense, the National Institutes of Health, the National Academy of Science, and the Department of Energy), does support basic (nonapplied) research. And even when the government is contracting for a specific product, research and development effort may be specifically reimbursed. In private industry the cost of R&D comes out of profit. One therefore often finds a higher level of conceptual work and technical sophistication in government-funded fields such as aerospace than in the nail business. One of the ironies in our capitalistic system in the United States is that the government now sponsors research in science and technology that is more technically advanced than the research sponsored by the private sector. Consequently, we have become innovators of all manner of high-tech components and systems but we find ourselves slipping in our ability to compete commercially with countries that focus on consumer goods. Our nation's ability to produce state-of-the-art weapons seems to exceed our ability to produce automobiles and consumer electronics that can capture the market.

Let us look in a bit more detail at the forces that cause organizations to take certain technological directions. Some people who have pondered this question have concluded that technology determines the desire for goods and services. They see technology and the knowledge that underpins it as the major driving force within a culture. People subscribing to this theory will sometimes say that technological "progress" is "autonomous" and speak of the "technological imperative." A perspective that often follows, sometimes called "technological determinism," assumes that technology determines social directions and that people will follow its lead.

A hint of this sort of outlook can be seen in Lynn White's excellent book *Medieval Technology and Social Change*.[4] White argues that the invention of the stirrup played a major role in changing the social structure of Europe. Without the stirrup, a rider was simply too unstable to fight from his horse. A missed swing of a large sword would result in the combatant being on the ground. Before the invention of the stirrup, the horse was used primarily for transportation to the scene of the fighting. The stirrup appears to have been first used in the East and imported to Europe in the first part of the eighth century. The Franks, under Charles Martel, realized that this simple addition to the saddle would make a whole new type of warfare possible. If the rider could hold a lance firmly to his body, the momentum of the horse would be added to his own and the lance would be unstoppable by the defenses available at that time.

Like most breakthroughs in military technology, this one led to escalation. The enemies of the Franks also adopted the horse, forcing the Franks to use bigger horses and increased armor for both rider and horse to maintain their advantage. Their enemies responded in kind, and soon large amounts of money were involved. In the early days of such combat one man's equipment cost about as much as twenty oxen and was equivalent to the cost of the equipment used by ten peasant families. In addition, equipment was needed for the knight's squire, and extra horses were needed to replace those killed in combat.

According to White and those who agree with him, feudalism resulted from the need for the resources to support this new type of warfare, which had in turn resulted from the technical breakthrough of the stirrup. A major resource in this time was land and the food it produced. Charles Martel and his heirs seized large amounts of Church land and distributed it to vassals on condition that they serve as knights. These vassals became the noble lords that oversaw the peasants. The mobile nature of mounted warfare resulted in fortified castles. Heraldry became necessary when armor became so all-concealing that knights were no longer identifiable, and tournaments became the means of practice for battle. Much of the social hierarchy, pageantry, and agony of the medieval period in Europe that we studied in our history books traces its existence back to the introduction of the stirrup in this scenario of technological determinism.

This notion that technological breakthroughs are the dominant force in altering the fabric of society may be overly simplistic, but many people subscribe to it. Certainly many people personally engaged in technology often believe that developments in technology should shape progress. This is understandable, since engineers who love their craft feel that everyone must, and that new technology is good. Such people often start new companies secure in the belief that a market for their new knowledge will exist. To some extent this belief is true. In the early days of the computer industry there seemed to be a market for most new technological advances, and the military can always be counted on as a bastion of technological optimism, as evidenced by the U.S. defense establishment's support of technology and science. The military has traditionally funded extremely advanced research and development in the belief that technological breakthroughs will bring new weapons, which in turn will make it less likely that our country will be attacked and, if attacked, will make it more likely that we will win. Ironically, technological development may have reached the point where the type of war that the military loves the most, with its traditional strategy, tactics, and rules of behavior, may no longer be feasible. Weapons of mass destruction nullify the warrior skills of the individual.

Historical counterexamples to the philosophy of technological determinism abound, however. Sophisticated technology does not always determine the outcome of war, as shown in Vietnam. Business lore is full of stories of technology-based companies that failed. In general, U.S. industries feel that this country is technologically more advanced than its counterparts around the Pacific rim. Yet the competition from these smaller nations is much fiercer than technological determinism should allow.

China, over the course of its long history, has produced a tremendous amount of technological invention, yet the Chinese have not pursued their innovations for economic gain nearly as vigorously as Western countries have. The Japanese, although they were early innovators in the development of firearms, elected not to use them for 300 years.[5] Something must have been at work in these cultures other than technological determinism. Could it be that societies determine the direction of technology, rather than the other way around? That the emotions of individuals and moods of a culture are the key?

The business world is fond of the theory that technology responds to the market by providing what people want. This is often referred to as "market pull," as opposed to "technology push." Certainly evidence can be found for this view. The supersonic transport has been a commercial failure simply because people don't want to cross the ocean that fast badly enough to pay the price. Many inventions have no effect, as can be seen by looking through the file of U.S. patents. This philosophy might be called customer determinism. Rather than technology's determining societal directions, technology responds to human desires and needs as defined by the market. People subscribing to this theory would assume that the stirrup was invented because warriors figured out that they needed some way to handle heavier weapons while mounted and were willing to pay for it.

However, this simple explanation also does not stand up to detailed examination. Nathan Rosenberg, in his book *Inside the Black Box: Technology and Economics,* reviews ten formal studies of technological innovations that are often cited as arguments supporting the dominance of marketing as the forcing function for successful technological innovation.[6] His conclusion, with which I agree, is that the studies are flawed and things are not that simple. Both technology and market are important. Technology rarely operates in isolation from utility. Technological capability and the market affect each other. Certainly people are attracted to unimagined technical breakthroughs, and certainly organizations create demand through advertising. But it is equally true that technology responds to existing markets. It is a version of the chicken and egg problem, not chicken *or* egg.

Organizations must undertake projects that are technologically within reach and for which there is a market. Smart businesses realize that successful products are both pushed by technological capability and pulled by market demand. Technological progress requires both the ability to accomplish it and the desire to buy it. Neither market demand nor technological ability alone is sufficient. There was good technology in the Edsel, in teddy bears containing integrated circuits that allowed them to talk, and in home video discs, but the market was not there. During the 1940s the public desired automobiles that would convert into airplanes and private helicopters, but available technology did not permit acceptable products for the market (it still does not).

Since technology is an activity of people, emotions play a critical role in determining the problems on which engineers work. Technology is often considered to be a completely rational enterprise, producing products for rational customers. Nothing could be further from the truth. Most people own cameras, compact disc players, automobiles, and other products that have features they do not use. Do they really take pictures at $1/1000$ of a second? Do they really program the tracks on their disc? Most personal computer owners I know are quite aware of the amount of memory in their computer, but they are not aware of how much of it they are using. If they found out, they would undoubtedly realize that their computer has a large amount of capability they do not use. Does that mean that the excess capability is not valuable? Not at all. It has value because they love it.

I was firmly convinced, during my more cynical periods at the Jet Propulsion Laboratory, that as far as the public was concerned, the entire U.S. space effort was entertainment. I still believe that the public supports large projects in part because of the pageantry. How about the Strategic Defense Initiative? Is it coincidence that we call it Star Wars, or is there a bit of Lucasfilm attraction there? How about airplanes? I am always running into my politically liberal antiwar university friends watching the Blue Angels at Armed Forces Day events or standing in line to watch movies like *Top Gun*.

The direction of technology involves desires, dislikes, hatreds, and passion. I spent my military time in the Air Force at the Air Force Flight Test Center at Edwards Air Force base. No one can convince me that policymakers in the military are not fascinated by technology to a point beyond the logical. I have seen too many of my lower-ranked friends struggling to write rational justi-

A watch from Audemars Piguet's CEO Collection. Automatic movement indicates the date, the day, the month, the chronometric time, as well as the phases of the moon. The watch is mechanically programmed to accommodate different month lengths and leap years until the year 2100. Technology push or market pull?

fications to demonstrate the need for a military gadget someone of higher rank loved. As far as I can tell from reading military history, the weapons that have made crucial differences in wars are technically advanced but dependable in battle and consistent with the skills of the users. Many of the weapons desired by the military are so on the edge of feasibility that they may not fulfill their role in battle. The military mind respects honor, glory, and warriors and seems genuinely reluctant to see such things as aerial dogfighting and face-to-face battlefield confrontations disappear, even though several conflicts have now demonstrated their inefficiency. The military is only one example of a set of customers that are highly motivated by emotion, even though they seldom admit it.

Engineers, too, are emotional, and this affects the directions taken by technology. As an engineer, I seek problems I like to work on, I am biased toward solutions that please me, and I like mechanical things. If I had participated in the early development of television, I would probably have tried to do it mechanically rather than electronically, which would have been unfortunate, since such things are best accomplished through electronics. I consciously and unconsciously try to manipulate my environment so that it produces more of the type of work I like, and probably have some success.

Many other forces influence the directions of technology, such as chance. James Burke's very popular TV series *Connections* views technology as a series of fortuitously interdependent events.[7] His connections have to do with the chance combinations of cultural needs, political alliances, economic desires, and timing that caused certain technical and scientific ideas to have powerful effects. This viewpoint explains, for instance, why da Vinci's invention of the parachute and the tank had no effect: The world was simply not ready for them. I am not as fatalistic in my personal viewpoint about technological progress as is Mr. Burke, but it is certainly true that luck plays a role.

Beyond technology push, market pull, and the emotional factors that go into both of these forces, a number of other influences on the activities of engineers and the problems they solve should be mentioned. Probably the first essential for technological innovation is an economy wealthy enough to allow energy and resources to be invested in improving the standard of living. Little technological development occurs when 100 percent of the economic resources of a culture must go toward maintaining life. Technology develops rapidly after something happens to raise people above a subsistence level. This occurred during the neolithic revolution, after the domestication of animals and agriculture made additional resources available, and during the Industrial Revolution, because Europe became wealthier through trade and colonization and had a wider range of raw materials to work with.

Technology also relies upon other technologies. Rosenberg writes about complementarities. The nature of agriculture (and therefore agricultural engineering) was changed drastically by low-cost transportation of farm products made possible by the railroad and the iron steamship. The fledgling electrical

A Patriot missile, developed by Raytheon Corporation, which came to the public's attention for its ability to intercept SCUD missiles during the Persian Gulf War.

industry was aided by simultaneous advances in the generation of electricity, the transmission of power, the ability to measure power, and electric lamps and motors.

Technology must be consistent with cultural views if it is to succeed. As an example, the United States is an extremely toilet-trained culture, and we therefore expect our sewage systems to be reasonably hidden from view. We also gave the world so-called disposable diapers, destined to occupy land-fill dumps for millennia.

Technology is also symbolic in our society. We want to be #1 in technological capability in the world. We were unable to accept Russia's prowess in space in the 1960s and put tremendous effort into attempting to gain the lead in the "space race." I remember constant attempts to rationalize our second-rate performance. One of my favorites was the argument that Russia had developed very large booster rockets because they were not technologically smart enough to miniaturize payloads. I thought this was particularly interesting, since it seemed to me that we were forced to miniaturize because we lacked the ability to develop adequate booster rockets. We still measure ourselves by technological accomplishment. The distress that followed the *Challenger* disaster and accompanies Japanese success in electronics not only reflects loss of life and property, but also damage to our self-image. The direction of technology in the United States is influenced by our technological hubris.

The role of our government in affecting the directions of technology must not be underestimated. In the United States the government first of all sponsors the development of new technological capability by funding R&D on specific projects in the private sector, by underwriting basic research and development by means of agencies such as the National Science Foundation and the National Institutes of Health, and by running its own laboratories such as the National Aeronautics and Space Administration Centers and the Naval Ordnance Test Station. Approximately one-half of all U.S. research and development is funded by the government (see table). Second, the government is a major source of procurement for the products engineers make. In this role, it buys everything from missiles to bedpans.

The costs and benefits of our government's role in technology have been

**Federal funding for research and development
for fiscal year 1989 (excluding R&D plant)**

Budget function	Millions of dollars
National defense	40,574
Health	7,724
Space research and technology	4,589
General science	2,379
Energy	2,427
Transportation	1,019
Natural resources and environment	1,208
Agriculture	910
Education, training, employment	343
International affairs	141
Veterans benefits and services	212
Commerce and housing credit	134
Administration of justice	45
Community and regional development	77
Total:	61,823

Source: *Statistical Abstract of the United States*, 1990, p. 585

hotly debated, particularly in areas such as defense and space. The government tends to justify such efforts in part as producing technological capability that benefits private citizens and economically strengthens the nation in foreign trade. As part of this justification, the government makes periodic studies attempting to trace research and development into the public sector. The other side of this argument is that this type of spending is not very effective in helping the private sector compete—that one of Japan's advantages is that its research and development efforts are not diverted to military problems.

Whatever the answer to this debate, the role of the government makes it even more difficult to create simple theories as to why technology proceeds in the directions it does. This is true in most countries. In certain nations having a stronger technology policy than the United States the government takes an even more central role in determining technological capability. We have read much about MITI in Japan and the active role it takes in technological development. In India the government takes an extremely active role in increasing technological development by establishing economic barriers so high it is possible for local companies to compete with sophisticated foreign companies by licensing technology and learning to use it as they apply it.

Are the forces presently causing technology to take the directions it does the right ones for the long-term good of humanity? The answer to this is not straightforward. In his book *The Culture of Technology,* Arnold Pacey discusses this topic in depth using a nomenclature that I like.[8] He refers to three sets of values that he claims control the directions of technology. One of these, which he calls the virtuosity values, has to do with the quality of technological capability—breakthroughs, sophistication, and accomplishments. The second set, economic values, has to do with technology as a force for economic growth. The third, user or need values, has to do with the extent to which technology serves the individual and improves the quality of life. It seems clear that at present in the wealthier countries the virtuosity and economic values dominate over the user or need values. We have talked about technology push and market pull, but not about things such as an adequate standard of living for all of the world's people or personal fulfillment. In addition, the virtuosity and economic values are based on assumptions of infinite resources and eternal growth. These assumptions are already being questioned and are clearly not valid forever. It is probable that the forces that direct technology will change over time, because the present forces that direct technology take little account of a finite Earth.

In summary, the process that launches engineers onto problems and projects seems to reflect the nature of technology itself. It is terribly human. It is a mixture of intellectual interest, greed, love, and chance. Certainly technology itself provides some momentum: It is a fascinating and rewarding game to play. Technology does "push," and when the capability to do something is available, the something often occurs, for better or worse. But technology traditionally has served people. To be successful, someone must desire its products. It is market "pulled" and responds to social needs, in which emotions play a large, but not always publicly, recognized role.

But there is also the element of chance. Phrases such as "technology policy," "technology planning," and "technology management" are quite new. We are just beginning to attempt to more logically control the progress of technology. Still, technological developments beget other developments and result in problems that are not generally predictable. During a project, engineers must continually respond to unexpected problems and difficulties, and final products are often quite different from those initially expected. The engineering process is neither completely predictable nor completely controllable. It is, in that sense, like all other aspects of life. Finally, the forces that presently cause technology, and therefore engineers, to work on particular problems are not valid for eternity. They are based on assumptions dating from a time when populations were smaller, technological capability was weaker, and the earth seemed infinite. As technology acquires the ability to have ever larger impacts, as populations grow, as resources dwindle, and as the pressure for equity between cultures increases, these assumptions will have to change. As they do, so will the problems on which engineers will focus.

4

Design and Invention

The Concept

What does the engineer do after a problem is defined? Typically he or she has to design or invent something. These activities are the stereotypical "creative" aspects of technology. This is an unfair stereotype in that all phases of engineering require creativity. However, it is in giving birth to the concept that we gain the interest of the press, the media, and the general public. Probably many of you recognize the illustration on page 79 as a sketch by Leonardo da Vinci, who among other things drew configurations for parachutes, helicopters, military tanks, and bicycles. We are tremendously impressed with the man and his ability to conceptualize products that would not be built for 400 years. These sketches probably had no effect on the world whatsoever, because not only were his notebooks lost for a long time, but when these concepts became real they evolved from different origins. Leonardo was so far ahead of technology that his ideas had no chance for implementation at the time. Nevertheless, we are impressed.

I have been interested in creativity for a long time. I study it, teach courses in it, and consult with all sorts of people interested in using and managing it better. I am always struck by how much value people give to the idea itself, even though producing the original concept is usually an easy task compared with the effort required to implement it well and sell it to a conservative world. We love ideas and venerate those who originally produced concepts that later affected our lives. In the case of Leonardo, we are not even concerned with whether his ideas were in any way responsible for the eventual development of the products.

I have served most of my engineering time in design. I have always loved the blank piece of paper and the drama of initially filling it. The people who think of concepts more brilliant than mine have always intrigued me. However, I am most impressed with concepts that find reality, with inventions and designs that reach physical form and fulfill their intended function successfully.

Our culture seems to be particularly fascinated with the notion of "inventors." We were brought up on the stories of Gutenberg, Watt, Edison, Marconi, the Wright Brothers, McCormick, and Whitney. We become annoyed when Russians claim to have invented things first, and we consider people from cultures that are less enthralled with invention than ours to be less

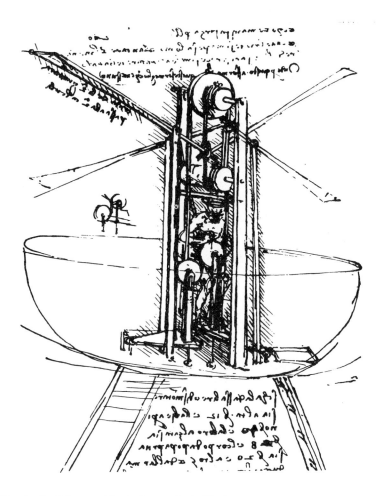

Sketch of a flying machine by Leonardo da Vinci.

creative, and by implication less smart. Japan has traditionally been less obsessed by invention than the United States. Are the Japanese less creative than we are? I think not. Theirs is a much older culture on much more limited land, and they have necessarily focused on doing existing things very well, in addition to inventing new things. They put more creativity into process and less into invention. At this point in history, that propensity is standing them in good stead.

People and history books have always had a desire to give an individual credit for technological inventions. In Chapter 1 I mentioned many names of such people. Yet detailed examinations show that a single person very rarely deserves total credit for an invention. Think of some of the great inventions of the past. How about the printing press? Did Johann Gutenberg (1400–1467) simply sit down and invent printing? He was a goldsmith who developed a printing office in Mainz, Germany, along with Johann Fust (1400–1466) and Peter Schoeffer (1425–1502), Fust's son-in-law. More than fifty books have been connected with this shop. However, printing would not have been commercially feasible without paper, and paper was introduced to Europe by the Islamic Empire early in the twelfth century. Block printing also apparently came from the East. Marco Polo had described the use of printed paper currency in Kublai Khan's empire in the thirteenth century, and Archbishop John of Monte Corvino described its use in China about the same time. Certainly by the fifteenth century block printing was widely used for making playing cards, religious images, and even books. Similarly, printing inks and presses were in use before Gutenberg's shop. The issue seems to come down to the question of who invented moveable type.

Here, too, the record does not give Gutenberg full credit. There are printed materials from Holland that supposedly predate the Mainz shop. Early work on moveable type in France was also under way. Gutenberg did invent a brilliant approach to the mold in which the type was made. His mold was reusable and gave the carefully controlled dimensions necessary to align the type with the proper spacing. His shop was also a major player in printing. The work on the famous Bible did not begin until approximately 1450, and by that time some sophistication was coming into the casting of type and setting it. Printing the Bible a page at a time would have required about 20,000 pieces of type.[1] Later stages of production produced six pages at a time, and some 100,000 pieces would have been needed. The production of this quantity of type would have taken two men a year. It would have taken an individual at least half a day to set a page of such primitive type, and apparently six presses were used to print the Bible. Other people would be needed to load the paper, ink the type, hang the printed sheets to dry, and so on. In 1450 the Gutenberg–Fust–Schoeffer shop probably employed about 25 people. This was certainly a reasonably significant enterprise for that time, but did Gutenberg really invent printing?

How about another example? Let's look at the later one—the steam engine. The precedents for steam power were many. As mentioned in Chapter 1, Hero of Alexandria built a turbine as a curiosity and a heat engine of sorts to open temple doors. People in the seventeenth century were interested in the ability of steam to force water into different locations and to produce a vacuum through condensation. Giovanni Branca (1571–1640), an Italian architect, described an impulse turbine that would use a jet of steam, although mechanical technology was not sufficiently advanced to build a working model. Toricelli, Pascal, and von Guericke did their famous work on vacuum in this century, and in 1690 Denis Papin built an engine which used steam to raise a piston to the top of a vertical cylinder, where it was held by a latch. After the steam had condensed in the tube, the latch was released and atmospheric pressure forced the piston down.

Thomas Savery used this principle to invent a steam pump, which saw quite a bit of application, since pumping water out of mines and getting water into large buildings were major goals at this time. A similar device was independently developed by Thomas Newcomen (1663–1729). Although a superior machine, Newcomen ran into Savery's patent, which broadly covered all applications of "the impellent force of fire." He therefore entered a partnership with Savery. Newcomen continued to improve his engine and by the early part of the eighteenth century it was in wide application for pumping. After Newcomen's death, his engine was still further improved by men such as John Smeaton, a famous eighteenth-century engineer. Not until the latter part of the century did James Watt enter the picture. He was a brilliant engineer who drastically improved the performance of the steam engine. The external condensor, which he invented, dramatically increased the engine's efficiency by making it unnecessary to cool the cylinder for each stroke. Watt also patented double action, which increases power by allowing the steam to push alternately on each side of the piston, and the concept of cutting off the steam supply before the end of the piston stroke in order to make better use of the expansion of the steam. In collaboration with Matthew Boulton, a very successful manufacturer, Watt raised the steam engine to new levels of efficiency and practicality, but did he invent it? No, and yet our schools are filled with children learning that he did.

It is somewhat ironic, given our fascination with "inventors," that when we attach names to technological inventions, we often (mistakenly?) choose someone associated with the implementation and successful dissemination of the idea, rather than the person who came up with the original design or concept. The first man who built a vehicle powered by an internal combustion petroleum-fueled engine was named Siegfried Marcus (1831–1899). However, we seem to remember people such as Gottlieb Daimler (1834–1900) and Karl Benz (1844–1929), since they are enshrined in factories that have made a large and visible impact. I occasionally meet people who think that Henry Ford invented the automobile. He is best known for the popularization of the motor car through low-cost production. But he did not even invent the assembly line. The automobile business was preceded by the meat-packing business, among other enterprises, which used the assembly-line concept.

How about more modern inventions: the jet engine? the computer? radar? gene splicing? the microprocessor? Here we are a bit hazy, aren't we? We don't have someone as convenient as a Cyrus Hall McCormick or an Alexander Graham Bell (both of whom were also dependent on previous work). Perhaps we are becoming more aware of the complexity of the team effort necessary to attain such sophisticated technological achievements, or maybe these inventions are simply too recent for historians to have yet adequately publicized a single inventor.

What is the present state of invention in the world? It is alive and well. How about the United States? Here there is room for some concern. In 1980 66,200 U.S. patents were issued. In 1988 there were 84,272, an obvious upward trend. However, approximately 47 percent of the 1988 patents went to foreign individuals and companies, as compared with 19 percent in 1963, 37 percent in 1980, 45 percent in 1986. Japan was issued by far the largest number (see table). Does this trend indicate trouble for the United States? Not necessarily. The absolute number of patents issued to U.S. individuals and corporations rose from 37,500 in 1980 to 47,700 in 1988. The United States is obviously losing its dominant position as inventor to the world, but this is due to an increase in inventiveness on the part of others rather than a decrease on our part.

U.S. patents issued to foreign individuals and companies in 1988

Country	Number of U.S. patents	Percent of total U.S. patents
Japan	16,984	20.2
West Germany	7,501	8.9
France	2,792	3.3
U.K.	2,758	3.3
Canada	1,642	1.9
Switzerland	1,298	1.5
Italy	1,172	1.4
Netherlands	899	1.1
Sweden	891	1.1
Taiwan	536	0.6
Other	3,231	3.8
All foreigners	39,702	47.1
Total	84,272	100.0

Source: *C & EN*, March 27, 1989

Of the U.S. patents issued, approximately 17 percent were issued to individuals, rather than to the government, who received 1 percent, or to corporations, who received the remaining 82 percent. This percentage to individuals was 20 percent in 1980, 21 percent in the 1970s, and 23 percent in the 1960s. This trend has also caused some concern based on the feeling that the individual inventor has been responsible for some of the great breakthroughs of history. I was once an educational consultant to a fascinating group called the National Inventor's Council, an advisory group to the Department of Commerce, which includes the patent office. The council was made up of two dozen highly prestigious inventors, such as Chester Carlson (xerography), John Bardeen (the transistor), William McClean (Sidewinder missile), J. Presper Eckart (computer), Charles Stark Draper (inertial guidance), and Jack

Rabinow (self-regulating watch and countless other devices). The group spent much of its time discussing the plight of independent inventors and possible changes in the laws to better motivate and support them. Yet even this obviously creative group could not think of much that would do so. For better or worse, inventors attempting to work by themselves or with a few assistants are at an increasing disadvantage, both because the process of applying for and protecting patents has become increasingly complex and costly and because as technology becomes more complex, larger amounts of capital and more numerous disciplines are needed to produce the parts and environments necessary to experiment with new concepts. For this reason, more and more invention is done formally in an organizational context, often under the rubric of research and development. Certain industry groups (aircraft and missiles, electrical equipment, machinery, chemicals, and transportation) spend the lion's share of the funding, but a good portion of R&D occurs outside of industry (see tables).

Invention may come not only from independent inventors and formalized research and development activities but also from clever ideas produced in conjunction with regular design, production, and operations. Historically, the designer has been asked to be ingenious, and organizations are aware of the advantages of applying for patents for ideas that might improve their position, whether to protect them against use by their competition or to sell or license them. What does this mean to the engineer?

Typically if one invents something while working for a profit-making organization, one's name is on the patent as inventor but the organization retains the right of ownership of the patent and the right to any resulting profits. Organizations often have a policy of giving the employee/inventor the rights if the patent does not have anything to do with the business—for instance, the invention of a toy by an engineer in charge of chassis and drive-train development in the automotive business. Some organizations may give the inventor a percentage of the profit (a royalty), as was occasionally done in the toy industry. Even more enlightened organizations, such as the one I work for, will give the inventor the rights to anything, provided that the work was not done for a funding agency that contractually demands the rights (such as the government).

Total funds (private plus government) for performance of industrial R&D by selected industries, 1980 and 1986, in millions of dollars

Industry	1980	1986
Aircraft and missiles	9,199	16,240
Electrical equipment	9,175	18,030
Machinery	5,901	10,696
Chemicals and allied products	4,636	9,021
Motor vehicles	4,956	(unavail.)
Professional and scientific instruments	3,029	5,421
Petroleum refining	1,522	(unavail.)
Other	6,059	(unavail.)
Total	44,506	80,629

Source: *Statistical Abstract of the United States*, 1990, p. 587

Scientists and engineers employed in R&D in 1987

Sector	Number (in thousands)
Industry	594.8
Universities and colleges	119.1
Scientists and engineers	87.9
Graduate students	31.2
Federal government	62.3
Other nonprofit institutions	30.0
Total	806.2

Source: *Statistical Abstract of the United States*, 1990, p. 588

Stanford University has an interesting, motivating, and successful program in patent licensing. If an employee or student invents something, it can be submitted to a group of people organized by the university who do a preliminary study of its profitability. If it looks as though it has profit-making potential, the university will patent and license the invention. One third of any resulting profits goes to the inventor, one third to the academic unit (department) of the inventor, and the final one third to the university. This is a very reasonable program in an institution in which people produce many ideas but are not particularly interested in the legal and business hassles necessary to turn these ideas into profit. A patent application itself requires money above and beyond the fee and the charge for the search. A good job of patenting an invention may require several overlapping patents, with accompanying attorney's fees. Even beyond this, a patent is only a license to protect an invention legally; it does not prevent others from copying the idea. If one applies for a patent, one must be willing to protect the right later in court if necessary, which typically requires a large amount of time and energy and the unpleasant opportunity (for many) of dealing with attorneys and courts. Negotiating agreements with licensees is also a complex activity. For inventors who do not like to do such things, a potential one third of something is usually preferable to all of what they would receive if they personally exploited their ideas.

As discussed in Chapter 3, successful inventions must fulfill a need and be sensitive to technical feasibility. Braille speedometers and anti-gravity machines will not get you very far. If the inventor is to profit, either through fame, income, or pride, the idea must be nicely produced and successfully marketed. But how about the production of that idea itself? What can we say about the mental processes required in generating an original idea?

You can best get a feeling for the process by inventing something yourself. If you have never done this sort of thing, give it a whirl. You don't need to invent a better integrated circuit or airplane engine. Find a few problems that you think can be solved by application of technology that is at your level of confidence. Anyone can think up a gadget, a widget, or a gizmo. Inventors whose products become successful usually find that the following things are necessary:

(1) A heightened sensitivity to the problem and to solutions to similar ones. Successful inventors that I know are extremely problem-sensitive. They are tuned in to the little inconveniences or hardships in life that can be addressed by the technology they know. They are also highly knowledgeable about what has been done in their area. People who invent mechanical devices, for instance, are very aware of existing devices and their components and principles. People who invent sensors are very knowledgeable about the physical phenomena they want to detect and the sensors that are on the market. Professional inventors do not invent in a vacuum, unless they self-impose one for a period in search of fresh ideas.

(2) Examination of more alternative solutions to an idea than one usually deals with. Inventors rely on the ability of the mind to produce a wide variety of ideas, given the proper knowledge, time, stimulation, and motivation. Invention is puzzle solving, and it is necessary to force oneself away from the traditional, the stereotypical, and the precedented. Many inventors have their personal techniques for generating ideas.

(3) The confidence to pursue ideas that are criticized. Individuals sometimes like fresh ideas. Societies hardly ever do. Reactions to a new idea tend to be critical, and successful inventors have to understand and deal with this. One's own mind can even be a harsh critic. It is fully capable of wondering why, for instance, if an idea is so good, someone else has not already thought of it.

(4) The resources (time, money, other people) necessary to reduce the idea to believable form. The invention process typically requires fabrication of models and prototypes and testing of new principles.

(5) Communication and sales ability. Since the world is not eager for new approaches, one must be able to convince it that a particular approach fills a need and can be economically produced.

Now let us talk about design. Design may require invention, and all inventions must be designed. Design always includes defining a product to the point where it can be manufactured economically. This is a central and critical activity of engineering, and a large percentage of all engineers participate in the design process. Design requires a broad spectrum of knowledge and skills, some of them technical but many of them quite "soft." Design

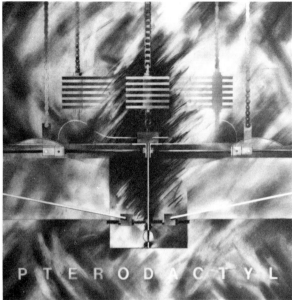

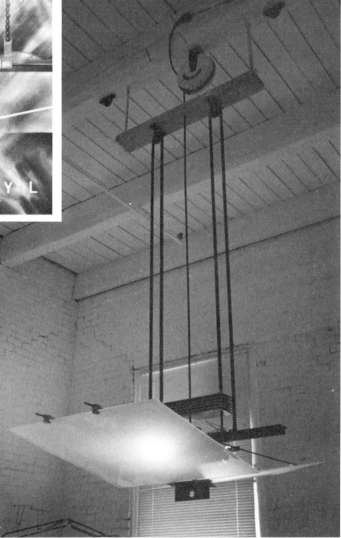

A hand-built prototype of a hanging light fixture, by Dirigo Design. The materials include sheets of sand-blasted tempered glass, lead counterweights, rectractable power cord, halogen lamps, and 30 feet of motorcycle chain. The rendering is by John Hopkins, one of the lamp's designers.

requires general problem solving, sophistication, and confidence. A great amount of art enters into design. The designer must be comfortable with technological experts, but there are no scientific techniques for determining the first lines to put on a blank piece of paper. The amount of science varies considerably. For instance, design of an integrated circuit relies heavily on theoretical understanding and is not possible without the aid of a computer. On the other hand, design of a mechanical device may draw principally on the designer's experience and if necessary could be done by hand (see figure). In all cases, however, design requires judgment and a sophisticated aesthetic sense. It requires a high level of creativity, and a head full of practical knowledge about the physical world.

Since some aspects of design are not analytical or scientific, the teaching of design does not fit well in prestigious engineering schools, whose faculty are usually more comfortable teaching mathematics and science than teaching judgment, aesthetics, creativity, and sensitivity to the physical world. Many professors of engineering wish that design were more of a science. Indeed, there are periodic movements in engineering to attempt to make design more analytical. Needless to say, they fail with the softer parts of design. Design is a much more critical and prestigious activity in practice than it is in engineering education.

As with invention, the best way to get a feeling for design is to design something. I am presently co-teaching a year-long sequence of courses on the nature of technology, science, and mathematics. My co-teachers are a mathematician and a physicist. The sequence is intended for students who are not majoring in engineering or science. Although they are bright, they generally feel only marginally competent at mathematics, science, and engineering. Fortunately, their self-perception is not entirely accurate. They know a great deal about these topics from their schooling, their general reading, and their experiences in life. Education being what it is, they just do not realize this.

During the first two segments of this sequence, the class looks at a large number of pieces of technology and the students try their hand at various exercises in the engineering disciplines. During the final segment, in order to show them more of the picture, I divide the students into groups and ask each

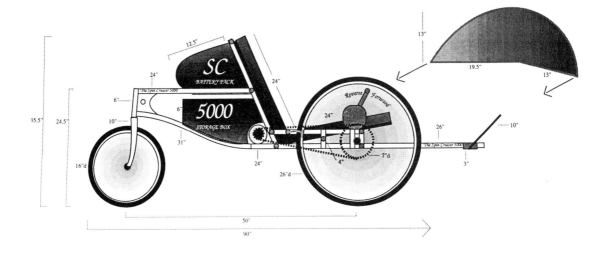

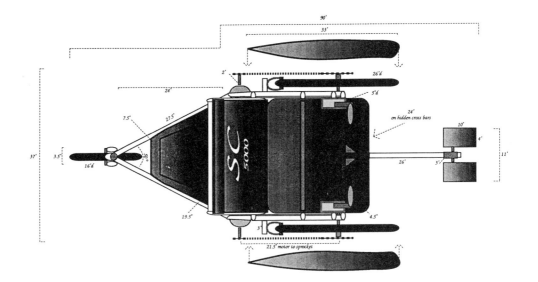

Students' design of an electrical vehicle.

group to design something. Their designs are on paper, but they must describe not only the product but each piece in enough detail that it could be made or purchased. The last time I did this, I asked them to design a small electrical vehicle for the customer of their choice.

Their first reaction was panic. Their response was, "We don't know how to do that." My response was, "Yes, you do." As is usually the case, they all succeeded beyond their wildest dreams. In order to increase the pressure, we even built one of them. It works beautifully. The adjacent figures show a design generated by one group on their trusty computer, and on page 93 is a photograph of the one that was built. In doing this project, the students had to perform the following intellectual feats:

(1) Become, as they say in business, "close to the customer." It is one thing to decide to design a small electrical vehicle for elderly invalids; it is another to gain the knowledge and understanding to do it well. Successful designers must be able to identify with the people who will use their products. Even if one designs products that are only indirectly used by people (such as unmanned spacecraft), one must be sensitive to the people who manufacture and assemble them, and the people who make the decisions about allocating resources.

(2) Learn more about that part of the physical world that is pertinent to the problem. For instance, one can learn a lot about small vehicles from studying bicycles, golf carts, lawnmowers, and other familiar machines. Most of the students in my class had never taken their bicycle apart and therefore were not sure how the handle-bar connected to the front fork in a way so that it all could pivot together. Needless to say, many local devices were disassembled during the project (most were successfully reassembled).

A sense of quality in engineering is essential, and comes from familiarity with products. In some areas of engineering if one has sufficient appreciation for and acquaintance with particular products, it is possible to do a good job of design without a great deal of technical education. I have an uncle, for instance, who became a machinist at a very early age and spent his career in a machine shop, rising to foreman and then to general manager. He became an extremely good machinist and always loved fine machinery, although he had no formal training as an engineer. Upon retirement, he designed and built specialty machines in his garage that were extraordinary in their sophistication and function.

One I remember clearly was a machine to make small lemon pies at very high speed. The machine would form pans from foil, mix the dough and filling, form the pie shell in the pan and inject the filling, cook the result, and then put whipped cream on top. To design a machine capable of passing Food and Drug Administration food-handling requirements (no noncleanable crevices, no corrosion) and capable of handling metal, pie dough, lemon filling, cooking, and whipped cream, all at high speed, was not something I would care to tackle. However, once while playing around he made a working steam engine with a piston approximately the size of a grain of rice, which I would also not want to have to do.

(3) Learn how to find out things they did not know. This necessitated looking through libraries and asking other people. A great amount of information used by engineers is in product catalogs, handbooks, and the heads of other people. If I want a ball bearing, I grab a ball bearing catalog and look up the number of one that matches my load, speed, and lifetime requirements. Bearing companies are smart enough to supply the world with their catalogs in order to sell bearings. If I can't find a catalog, I telephone the bearing company, which thoughtfully supplies sales and applications engineers to help me. I have reached a stage of efficiency (laziness?) in life where, if I need to know something, the first thing I do is ask someone who may know it. It is much quicker than heading for a library. I also like people better than libraries. Through this process, students learned that designers do not simply rely on what they already know. They go find out.

(4) Use a bit of "theory." The more facile the designer is with engineering theory the better, because design requires selecting materials, sizing parts, and optimizing function. The students in my design groups had to figure out how powerful a motor they would need for their desired performance and how much battery capacity would be required. They had to estimate how thick the frame members should be to support the load. In such situations, mathematics is essential. Much to their amazement, they had enough mathematical sophistication to handle the theory.

A student-designed electrical vehicle that works!

(5) Work with complicated geometries and complexity. Aids such as computers, drawings, and models are necessary. In addition, examination of options and integration of the final result requires juggling the product in the mind.

(6) Communicate—not only to interact with others during the design process but to communicate the results to others. This requires facility with graphic and three-dimensional modes of communications as well as with words and numerical relationships. Designers who are good communicators have tremendous influence over the direction of products.

During my early years at the Jet Propulsion Laboratory, we had no precedent for spacecraft. Buck Rogers and the people who drew covers for science fiction magazines were misleading, rather than helpful. It took effort to realize that spacecraft did not have to be red, pointed at the top, and equipped with wings. JPL's early spacecraft were designed by groups of people representing the various technical disciplines and perspectives desired. The physical requirements and opinions of the various members of the team were integrated by an extraordinarily talented senior designer, who worked at an old-fashioned drawing board. If the communication engineer wanted a 10-foot diameter parabolic antenna that would always point at Earth, the designer would think for a while and then draw one beautifully on the evolving spacecraft config-uration. All of a sudden, the antenna was real. The project and systems engineers would gather around his board and say things like "that looks good," or "the antenna is going to shade the solar panels at Venus encounter." The designer would keep iterating until finally everything fit together efficiently, using the input from the various specialists. In doing so, he determined the physical form of the spacecraft, and since these early spacecraft became a precedent, he contributed a great deal to the nature of present interplanetary spacecraft. I have worked with many design groups where a similar thing happened—a designer who is a good communicator is able to dominate the engineering process because of the many decisions that must be made that cannot be based on theory.

If you have the time and interest, try to design something. Make it simple, exploit other people as much as you can, and look for similar things in the world to give you ideas and details. Work in whatever medium you are comfortable with (paper surely, but also models).

The design process as it is done professionally begins with the development of specifications, budgets, schedules, and people assignments for the desired product. These result from the sorts of considerations discussed in the last chapter. The specifications are the description of the desired characteristics of the product and should be based on sound thinking about technical feasi-bility, the market, and so on. They may be so informal as to not require writing ("Hey Jim, can you figure out some way to keep this stupid engine from getting so hot?"). On the other side of the spectrum, they may be so complex as to require a small volume of print and pictures. They may also require a large amount of detailed analysis and experimentation. The speci-fications defining the limits on the dimensions of the passenger compartment of an automobile, for instance, are not only quite specific but are based on years of experience and intensive study of human anatomy and nervous-system performance. Changes in them require a good bit of work to ensure that the proposed new ones are acceptable. The specifications for a new airplane must take account of quantities as diverse as physical interfaces of engines and other purchased equipment, FAA structural and safety require-ments, airport characteristics, and existing maintenance capability as well as

Some considerations that may be addressed in design specifications

weight
speed, acceleration
control (response time, transfer function, input-output ratios)
cost
strength
dynamic behavior
chemical parameters (purity, corrosiveness, pH)
surface finish
electromagnetic parameters (watts of radiated power, voltage, current, resistance,
   strength of magnetic field)
thermal properties (temperature, thermal capacity, emmisivity)
reliability
lifetime
physical dimensions
physical tolerances
material description and properties
parameters having to do with compatibility with humans
safety
environmental criteria

the desired performance of the airplane itself. Not all of these quantities that may be addressed in design specifications (see list) are equally important, of course, and negotiation must take place in setting priorities. This negotiation (sometimes referred to as "making trade-offs") is an integral part of the design process.

After the specifications are set, design then proceeds into the stage of concept development. This is sometimes called "preliminary design" and may involve the investigation of one or a number of concepts to the point where they can be compared. This process includes calculations, experimentation, computer simulations, drawings, models, and many types of people. The specific nature of design varies with the product and the organization. The chemical engineering process may begin in the chemical laboratory, progress through a pilot plant to establish large-scale production parameters, and end with the construction of a plant. Computer software design may be primarily done on the computer before finally replicating the code on a disc. The design of a new piece of luggage may require a very large number of mock-ups to test the visual appeal of various shapes, colors, and textures. The role of the engineer in preliminary design varies accordingly.

In the case of hardware products, some people may be professional "board" designers. This term is an anachronism, since most designers these days work with a computer-aided design (CAD) system. On such systems one still works graphically, but the information from the "pencil" is stored in the computer, and can be used to produce drawings. In cases of highly automated manufacturing, the design information may be fed directly to a machine that makes the device. There are many ways in which design information may be used directly. For instance, in the manufacture of printed circuit boards for electronic assembly it is used to make a precise image that is the starting point for a process requiring photography and chemical etching (see Chapter 8).

In any case board designers, who may be engineers or highly trained "draftsmen" (a term doubly out of date because these people no longer draft and many are not men), are specialists in using visual representation to produce layouts of designs. A layout is a drawing showing the complete design, with all parts depicted as they will finally look. It does not contain the detailed information necessary to manufacture the part, but it shows the shape of each part and how the parts interact with the whole. Preliminary design often begins very informally, as the following figures show. They are drawings by Dennis Boyle, a designer for David Kelley Design, who was working on a small computer which, among other features, was to have a hinged cover that contained the display. This in turn required a mechanism to allow the cover to open and remain in any position, and a means of accommodating approximately 30 electrical conductors that necessarily connected the main body of the computer to the cover/display. Boyle, like many product designers, began by sketching ideas in a notebook, with a pencil. Some product designers begin on the CAD system, but most still prefer to begin with informal pencil sketches in order not to be constrained by the computer. During his preliminary thinking, he sketched several ways of dealing with the electrical conductors,

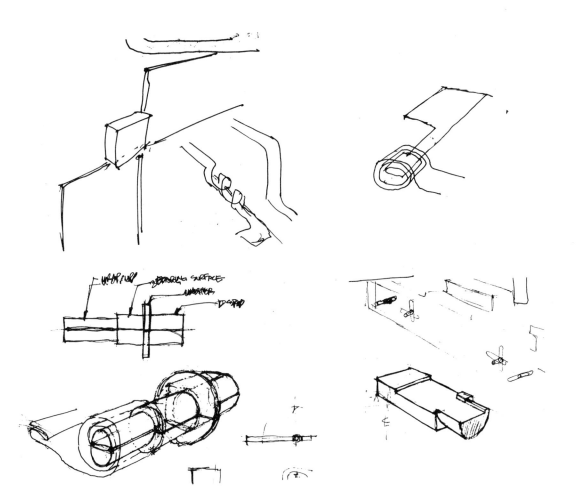

Dennis Boyle's preliminary design sketches of a hinged computer display screen.

which had to survive many openings of the cover. He first thought of using a flexible printed conductor that would essentially unroll and roll; he developed this concept and its hardware with increasing detail, and finally became quite specific on the bushing that was to clamp the printed conductor (see figure). The figure on page 98 shows an overall sketch of the mechanism and a rather well-developed concept of a torsion spring that would counterbalance the cover so that it would remain in any position and a friction brake to damp out oscillations. The illustration on page 99 shows a portion of the patent submission drawing covering the finished mechanism, and some of the details in perspective.

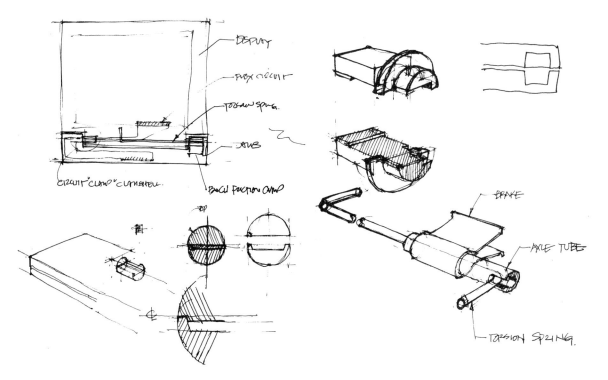

Boyle's design sketches.

Many people seem to think that engineers all sit at drafting boards and produce the type of formal drawings often called blueprints (although blue-printing as a process has not been used in engineering for years). Most engineers do not sit at drawing boards; they begin the design with informal sketches. Boyle's sketches show a progression from the type of squiggles produced to aid the thinking process in the early conceptual stage to drawings that communicate the design in detail.

Like specifications, the preliminary design process may be very informal. (I might not even tell my boss how I was going to keep his stupid engine cool; I might just go ahead and do it.) But for large and expensive systems (aircraft, automobiles), formal design reviews are held periodically, during which the design is presented to a panel of experts who criticize it and approve

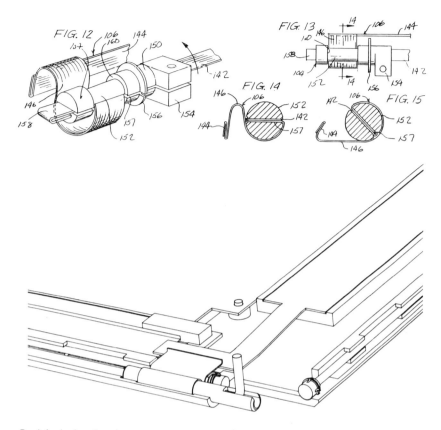

Boyle's design drawings.

it for the next stage of work or send it back to the drawing board. Once the final configuration is officially approved, design continues to the next stage: detailed (sometimes called detail) design.

In the design of products such as microcircuits and computer software, the preliminary and detailed design phases are integrated, since feasibility depends so closely on details and the computer is capable of handling both. In fact, even though preliminary and detailed design are often separately listed and scheduled, and may be done by different people, they are interrelated functions; problems can arise if those involved in preliminary design merely hand their work over to others and begin another project.

In detailed design, the product is reduced to complete descriptions so that it can be manufactured from the information produced by the designer. In the

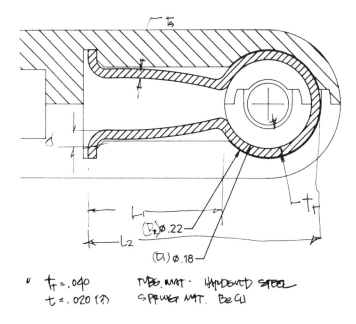

t_3

$(D_2) \emptyset.22$

$(D_1) \emptyset.18$

L_1

L_2

t_T

" $t_T = .040$ TUBE. MAT · HARDENED STEEL

$t = .020$ (?)   SPRING MAT. Be Cu

Boyle's detailed sketch of a friction brake spring.

case of hardware, this stage of design results in the production of a series of drawings. The first may be the layout. The drawing by Boyle reproduced above shows his friction brake spring in enough detail that one can see its final configuration. There are even a couple of dimensions and a material note or two. The next step in the process is the large, formal, detailed production drawing that will be used to manufacture the part.

A drawing, or set of drawings (if the part is extremely complex or if multiple processes are required for its manufacture), is produced for each part. In addition, special holding devices, tools, and even machines may be necessary for the manufacture of parts. In such instances drawings will be produced to describe these.

A detail drawing contains all of the information necessary to make the part. It specifies the material and its treatment, and, although it need not always be full-size, and not always even to scale, it shows the physical configuration of the part. The drawing also contains dimensions that describe the geometry of the part and tolerances on these dimensions. The dimensions, like the shape of the part itself, must be drawn in a way that is consistent with the manufacturing process. Square holes are not as much fun as round holes. Neither is a dimension that is based on an edge that will be cut away before the dimension is used.

The setting of tolerances is a critical part of detailed hardware design. The more accurate a part, the more it costs. It simply can never be exact. These tolerances tell the person who actually makes the part how much leeway is possible in the dimensions. The person who manufactures the part must remain within the tolerances or the part is unacceptable. The looser the tolerances can be, the cheaper the part will be to make. Most parts interact with other parts and with the world, however, and if the tolerances are too loose, the part will not fulfill its function. The proper choice of dimensions and tolerances requires a good understanding of manufacturing processes. More will be made of this in Chapter 8.

Detail drawings are standardized so that everyone can easily read them. The parts are shown in views (top, front, right side, and so on), and the views are related to one another in a particular way (in the United States the top view is above the front view, the right-side view to the right). Once one is used to this designation, drawings become quite easy to read and give one an excellent three-dimensional feel for the part. Much of the complexity is due to the dimensions and tolerances on the drawing. These are also done in a standardized way that is consistent with the manufacturing process. The designer therefore has to have a good sense of this process.

The drawing also contains the scale, helpful notes, final finish, and other information of value to those who will manufacture and handle the part. Finally, it contains information necessary in the future handling of the drawing itself. Each drawing has at least a number (so it can be found in the future, since drawings are kept as long as they can be of interest), the names of the part and the people who designed it, and a provision so that future changes in the drawing are recorded, along with identification of who made the change and the date of the change. This and other information pertinent to the organization producing the drawing can be found in the title block.

Assembly drawings may also be required. They tell how the parts are to be assembled into the final product. As before, special jigs and fixtures (holding devices) and other equipment may be necessary in assembly. Just as the detailed design of parts requires knowledge of manufacturing, the production of assembly drawings and assembly equipment requires knowledge of the assembly process. A drawing implying that parts should be assembled in an impossible or merely awkward sequence does a disservice and costs money. Most people who have tried to assemble a device bought at a store, such as a rowing machine, an outdoor grill, or a tricycle, are aware of the importance of good assembly drawings.

How does the designer acquire the necessary knowledge of manufacture and assembly? When products are simple and numbers of people are small, the knowledge often resides in the heads of the designers. In larger organizations producing complicated products, however, design is often done by a team of people, including people from production. This team will participate in the design from the beginning, when considerations of production should be introduced.

A design team will ideally include people representing all of the interests necessary to define the product completely. With an electronic instrument, for instance, this may include several types of electrical engineers (digital hardware and software, analog and control hardware, and power), mechanical engineers (mechanisms, heat transfer, perhaps fluids, hardware design), manufacturing and assembly engineers, industrial designers, representatives from related functions such as marketing, and a project engineer, who has final responsibility for the effort. The person in charge must not only have experience in designing complicated projects but must be comfortable in dealing with experts in the many disciplines, and be an inspirational leader, since good designers often have a lot of pride and almost all projects go through frustrating periods. The team will be supported by specialists in pertinent disciplines and functions.

Such teams vary in size from small (5 or 6 people) to rather large, as with the design of a large ship or a refinery. The team must be large enough to include the necessary knowledge and skills but small enough to take advantage of the high quality of communication, creativity, and motivation found in small work groups. The design team is, strangely enough, a recent development in some industries. Up until the 1970s, engineering often proceeded in a serial way, with the various engineering disciplines completing their portions of a design rather independently and then attempting to integrate them. After the dust died down from that rather traumatic exercise, the "completed" design would then be passed to the manufacturing group and more dust would get kicked up. This approach to design was organizationally convenient, because people were not expected to communicate with those in other disciplines, but it resulted in poorly integrated projects that were costly and difficult to produce. The United States was able to live with this handicap until the 1970s, when it began to suffer competitively at the hands of countries with a more interdisciplinary approach to design. The design team is now becoming the standard approach to engineering design, and I for one am delighted. To me, there is no more rewarding job than being a part of a motivated multidisciplinary design team working on a challenging and important product.

A quasi-discipline called "systems engineering" is oriented toward the integration of the various engineering functions and disciplines as well as the design of complex products in large organizations. The "systems approach" to engineering became popular during and after World War II, when the problem of organizing complex combinations of technical elements became extremely significant. When I was in the Air Force in the 1950s, great effort was being made to refer to "weapon systems," instead of to airplanes, starting carts, fuel tanks, and so on. The intent was to encourage people involved in complicated projects to think more about the integration of the parts in order to optimize the whole. Since that time, the systems approach has become integral to engineering.

Systems engineering includes activities that range from applying quantitative optimization and integration techniques to facilitating groups of people to aid them in reaching agreement. A few examples of activities essential to systems engineering are:

(1) Scheduling, predicting, and prioritizing. With a complex project, great attention must be paid to scheduling each effort so that the efforts integrate in a timely fashion. Many techniques are used to do this, often based on computers and often incorporating analysis of so-called critical paths, or

sequences of tasks that are particularly important in accomplishing a desired schedule.

(2) Defining interfaces. Since individual groups work on individual components, assemblies, or subsystems, they must each be given a specific set of constraints with which to work. As a basic example, if a group of people is designing a bumper that bolts to a frame being designed by another group, each group must be given certain dimensions having to do with bolt holes and such so that the two items can be fastened together. In the case of complex projects, the interface between two parts of the system may contain mechanical constraints having to do with geometry, movement, heat and fluid flow, and strength; electrical constraints having to do with current, voltage, and signal properties; chemical constraints having to do with reaction; and so on. The less restricting the interface definition can be, the more freedom the groups have to design. But if the definitions are too sparse, the need for ongoing communication and coordination increases, as does the risk of error. This becomes more critical if the groups are in different companies or from vastly dissimilar disciplines. The proper defining of interfaces is an art based on experience and on the ability to appreciate the nature of each group's task.

(3) Making trade-offs. In the design of a system, the optimization of one part of the system will usually handicap another. It is important that the system be "balanced," in that the overall system function be optimized. A drawing that was once endemic to the aircraft business showed an optimal system as seen by various groups that design pieces of the system. The aircraft, as optimized by the propulsion group, had huge engines. The communications group covered it with antennae and fiberglass body panels. The aerodynamics group had a fuselage with a tiny cross section and a huge wing. And so on. None of these special interests got it quite right. The making of trade-offs can often be done parametrically, which means that the variation of certain parameters with respect to others can be used to find an optimum point. In most complicated situations, quantitative thinking is required, but the final decision entails judgments based on values. Factors such as quality, risk, customer perception, and employee motivation are part of most trade-off decisions.

(4) Applying quantitative techniques. These are mathematical techniques for reaching optimal solutions in complex situations. A field of mathematics called operations research is devoted to this, as are the decision sciences in business schools. These techniques include linear programming, decision theory based on values and probabilities, game theory, and computer-based simulations.

(5) Communicating with people from a large variety of disciplines, both technical and nontechnical. Specialties each have their own basic principles, their own approaches to solving problems, their own techniques for analysis and synthesis, and their own jargon. At the engineering level, many people find themselves confronting disciplines that they thought they had avoided in school. I did not like electrical engineering as a student. At JPL I supervised a group that worried about the mechanical design of electrical equipment. I also worked on communication antennas and remote control and coexisted with people who talked of decibels, phase lag, back lobes, and the imaginary plane. Later I spent time as a systems engineer and had to deal with *all* the disciplines.

Now let us take a look at mathematics, perhaps the most important of the engineer's tools. Contrary to the impression gained by some engineering students, engineering is not all mathematics. Without mathematics, though, we would never have left the rule-of-thumb approaches of the Romans.

5

Mathematics

The Numerical Mystique

Mathematics is responsible both for the remarkable sophistication of our technology and for the discomfort many people feel about it. People who simply do not like mathematics often avoid learning about technology in order to make sure they do not have to do math. But one does not need to be a mathematician to understand the utility of mathematics in engineering.

Math enters the picture because engineers are asked to do such things as design a bridge that won't fall down, using a minimum of material. It would be prohibitively expensive, as well as dangerous, to hit a happy medium between safety and low cost with instinct and guesswork alone. If airplanes are not built extremely efficiently, they will not even fly. It is impossible to escape the laws of physics—and the symbols which express those relationships—when solving these sorts of problems.

It is unfortunate that so many people in the United States have negative feelings about mathematics. Math blocks begin early in life, when kids do not do as well as they might like in mathematics. This can happen for a number of reasons. People do have different aptitudes early in life. If they are slower at subtraction, or whatever, they can form an early dislike. Math blocks can also be a result of bad teaching or frustrations from poor test performance due to inadequate study. They can occur because of overly high expectations from parents or teachers, unusual abilities in other students, or social values that equate having mathematical ability with being a nerd. Whatever the reason, negative attitudes toward mathematics develop and deprive people of pleasure, vocational ability, and an understanding of engineering and science.

We all know something of arithmetic, one form of mathematics taught in grade school. Yet few people use it as much as they might in their thinking. Engineers do so regularly. It is a habit they acquire early in their careers, and a useful one that anyone can learn.

One useful application of quantitative thinking is to gain perspective on the very large and very small quantities involved in technology. Computer people speak of nanoseconds and picoseconds (billionths and trillionths of a second), and ship designers talk in thousands of tons and billions of dollars. Such numbers do not match our day-to-day experiences, and it is useful to bring them into some form in which they are meaningful to us. As an example, I cannot think about the national debt until I reduce the numbers to something I am familiar with. A national debt of over $2 trillion escapes me. Now I happen to know for some reason that a piece of typing paper is about .003 inches thick, so a dollar bill must be about the same. Using arithmetic, I figure that $2.5 trillion would make a stack of dollar bills about 7 billion inches tall. This is about 100,000 miles, or 12 times the diameter of the earth (which I also happen to know is about 8,000 miles). So $2.5 trillion is equal to 12 stacks of dollar bills, each equal in height to the diameter of the earth, which is something that I can at least begin to think about.

Since there are approximately 250 million people in the United States, each person's share of our national debt is about $10,000. However, 250 million is also difficult to think about. The Stanford football stadium holds about 80,000 people. I have seen it full, so I know how many people that is. If only 80,000 people assumed the national debt, each would owe over $30,000,000 dollars, a bit steep even for Stanford alumni. If everyone in the world assumed the U.S. national debt, on the other hand, each person would owe about $500, but I have even less feeling for 5 billion (the population of the world) than for 250 million.

Playing around like this helps me understand large and small quantities. In the case of the national debt I have trouble grounding it on any quantity I have a feeling for because of its overall magnitude. This is undoubtedly one of the reasons why it keeps growing; it is so big that no one understands what it means. A 20,000-ton ship is easier to grasp. It weighs about as much as 20,000 small cars (approximately 140 rows of 140). Or, if you prefer, three times the weight of the people who would fill Stanford stadium.

Quantitative thinking can quickly tell one whether technical things are feasible or not. As an example, many people are attracted to the idea of solar-powered automobiles. But discussions about their feasibility are endless if there is no quantitative content. Why are there not solar-powered cars on the street? The total amount of solar energy hitting the earth's surfce is about one kilowatt per square meter. That is simply all there is. A kilowatt is about 1.3 horsepower. A solar collector the size of a car (approximately 12 square meters), if aimed directly at the sun, would therefore collect about 16 horse-power. Unfortunately, conversion efficiencies are low, say 10 percent for photovoltaic (solar) cells. This means that 1.6 horsepower would be available. Given enough gears, that much power could move a car, but it wouldn't give you the performance of your 160-horsepower internal combustion automobile engine.

Could we collect solar power and store it in batteries? The only reasonably cheap high-capacity batteries presently available are the familiar ones already in cars—lead acid batteries. By looking in the Sears catalogue, I find that one of these typically holds about 90 ampere hours of energy, weighs a little under 50 pounds, occupies one third of a cubic foot, and costs $80. By looking up the energy content of gasoline in one of the many dull-looking books in my office, I find that a cubic foot of gasoline (8 gallons) contains about 25 times as much available energy as a cubic foot of battery and weighs about one third as much. A tank full of gasoline (16 gallons) therefore contains as much energy as 150 batteries, which would weigh approximately 7,500 lbs, make a pile 2 ft by 5 ft by 10 ft, and have an initial cost of $12,000. That makes gasoline look pretty good, doesn't it?

But does it rule out solar power for cars? No, if we could live with less performance and range in our automobiles, or if solar electric cars are a component of an overall system utilizing solar energy. John Reuyl, President of Energy Self-Reliance, Inc., designed an interesting system (U.S. Patent No. 4,182,290)[1] which combines a house with photovoltaic panels and batteries with a battery-powered electric car containing a small gasoline engine/generator. One configuration studied incorporated a residence with 110 square meters of panel, 61 kilowatt hours (kwh) of battery, a 10 kw DC/AC converter, and an automobile equipped with 31 kwh (680 kg or 1,500 lbs) of batteries, a 40 hp/16w electric drive motor, and a 25 hp/10kw engine/generator. Such a vehicle would have a 60-mile range, acceleration superior to many conventional automobiles, and a top speed considerably in excess of the legal limit.

In one mode of operation, the engine/generator would be used to make up the deficit between the energy supplied by the panels and that needed by the residence and automobile. In the other, the system would be connected to a commercial electrical power grid, so that excess electricity produced during the day could be put into the grid and power could be taken out during the night (public utilities are obliged to pay for any electricity you put into their system). Computer simulation indicated that such a system in California, if operating in the first mode, would require some 560 gallons of gasoline per year (assuming location in Fresno, and approximately 60 miles of automobile travel per day). In the second mode, some 900 kwh per year would be required from the utility. These energy needs are obviously much lower than that presently required to operate a residence and car and could be decreased still further by increasing the size of the panel and residence storage and decreasing the performance of the automobile. These figures were based on presently obtainable battery and solar-cell performance; improvements in these components could further improve the figures.

The study that produced these figures (and many others) represents an application of mathematics to investigate the feasibility of various approaches to the use of solar energy. The time and energy required were obviously very much less than building a number of different residences and automobiles in order to find the desired information. The solar-car argument can go on forever, but by using quantitative thinking, it can have some basis in reality.

Would you like to try such thinking? Here is a problem I gave my class last year. I asked them to do it quickly in their head, rather than trying to find the answer somewhere. Give it a try. One student's approach is in the note at the end of this chapter.

PROBLEM

As you know, certain parts of the U.S. are beginning to worry about shortages of fresh water. The daily fresh water usage in the U.S. in 1980 was as follows

Irrigation:	150 billion gallons
Public water utilities:	34 billion gallons
Rural domestic:	5 billion gallons
Industrial and misc.:	45 billion gallons
Electric utilities:	210 billion gallons
Total:	450 billion gallons

How much fresh water is used in the United States each day to flush toilets? Is this a significant problem to worry about?

Engineers usually work to greater accuracy than we have been doing here, and often to extreme accuracy (as when calculating launch trajectories for vehicles that will investigate other planets, or cross-sections for wing-spans in airplanes). Even approximate back-of-the-envelope approaches, however, can establish magnitudes.

Engineers use many forms of mathematics other than arithmetic, including algebra, geometry, trigonometry, calculus, boolean algebra, and statistics. Algebra, the first type of math most of us see after arithmetic, is extraordinarily powerful. It is tragic that so many people's math careers are ended at the point where general symbols such as x, y, and z are substituted for specific numbers. Algebra allows one to solve equations without being confused by words and to find relationships that hold true for any numbers. If you hated word problems in school, you perhaps did not realize that algebra makes them simple.

Have you ever been on an airplane in which the flight attendant moved people to the rear for takeoff? That is because the center of mass (or center of gravity, if you prefer) must stay within certain limits for good aerodynamic characteristics. Such moves are usually based on estimates of baggage and passenger weight. If we knew the weight on each set of wheels for the airplane, could we directly calculate the longitudinal location of the center of mass?

The answer is yes, because the laws of static equilibrium tell us that the total weight on the main wheels multiplied by the longitudinal distance from the line of the main wheels to the center of mass must be equal to the weight on the front wheels multiplied by the longitudinal distance from their point of contact to the center of mass (see figure). We can use this to set up a little algebraic equation. Let:

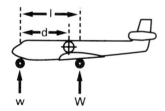

Weight distribution on airplane wheels.

W be the total weight on the main wheels

w be the weight on the nose wheel

d be the longitudinal distance from the point of contact of the nose wheel to the center of mass

l be the longitudinal distance from the point of contact of the nose wheels to the line of contact of the main wheels.

Using our equilibrium principle, we can now write:

$$wd = W(l - d)$$

which can be written $d(w + W) = Wl$ or $d = Wl/(w + W)$. This gives us a nice relationship for the longitudinal distance from the point of contact of the front wheels to the center of mass, since we know w, W, and l for our airplane. Better yet, it works for any airplane with a tricycle landing gear configuration.

Engineers cannot escape geometry, since the manufactured world consists of rectangles, circles, cylinders, parallelograms, cones, and other such familiar shapes. I remember participating in a design project a few years back in which a large flat panel had to be stowed in one location and then deployed to a completely different position and orientation. We were struggling with horribly complicated linkages consisting of many pivots, slides, and bars when one person in the group gleefully remembered from his high school geometry that a single simple hinge axis must exist somewhere in space about which a plane figure can be pivoted from an initial position to any other position. It took us all some time to remember how to find such an axis, but we succeeded because fortunately a member of the group had a son who had just completed high school geometry.

Triangles are often found in technology. If one has a number of pins and sticks with holes in the end and starts making two-dimensional structures, an interesting thing happens. Two sticks do not allow you to do much of value. Three sticks give you a triangle, a structure that has a definite shape. Four or more sticks give you structures that do not have a definite shape. For this reason, if you look at simple bridges, building frames, and other structures, you will see triangles. Since trigonometry is the study of triangles, it is an extremely useful type of mathematics.

Last year a student came to me with a tale of woe and sought my engineering wisdom. His dormitory had thrown a party and for novelty had decided to hang a keg of beer from a rope tied between two first-story window frames on opposite sides of the courtyard. They encountered difficulty in lifting it as high as they wanted, and therefore used a device called a come-along to tighten the rope. A come-along is a small winch capable of exerting a large

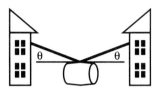

Keg on a rope.

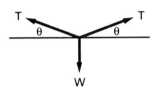

Keg in space.

force. The result of all of this was embarrassment because one of the window frames was torn out of the wall. The student wanted to know why.

Let us assume that the keg is in the middle of the rope, so that the angles are equal (see figure). Another rule of physical equilibrium tells us that the weight of the keg must be equal to the vertical component of the force on the rope. The tension T in the rope is pulling in both directions at an angle θ with the horizontal (see sketch). This force can be resolved into two components, one pulling vertically and one horizontally. The horizontal components simply cancel out by pulling against each other. The vertical component supports the weight of the keg (w). From trigonometry we know that the vertical component is equal to

$$2T\sin\theta$$

If you haven't taken trigonometry, $\sin\theta$ is shorthand for the sine of θ and is a unique number for any angle θ. If θ were an acute angle in a right triangle, the sine of θ is the ratio of the length of the side opposite to the angle to the length of the hypotenuse, or longest side of the triangle.

From our equilibrium rule, we can now write

$$W = 2T\sin\theta, \text{ or } T = W/2\sin\theta$$

If θ were 90 degrees, which would correspond to the doubled rope hanging straight down, the tension in it would be one half the weight of the keg, since the sine of 90 degrees is one. For an angle θ of 45 degrees, the sine is approximately 0.7, so that the tension in the rope is about ¾ of the weight of the keg. We eventually get into trouble, however. The sine of 10 degrees is 0.17, so that the tension in the rope would be almost 3 times the weight of the keg. At 5 degrees the tension would be about 6 times the weight of the keg. At 2 degrees, 14 times the weight of the keg. Finally, the sine of 0 degrees is zero, so that an attempt to take all sag out of the rope would result in infinite tension. Since this tension was being exerted against the window frame, and since kegs of beer are heavy, the students' problem was not surprising, merely expensive, since they had to pay to repair the window. An engineer would have hung the keg from second-story window frames and allowed the rope to sag.

Calculus enables the engineer to work with continuously varying quantities. Knowing the relationship between variables, rates of change can be found by differential calculus. Areas and volumes can be calculated with integral calculus. Equations containing not only constant quantities but also varying ones (differential equations) can be solved using calculus. It is of great value in predicting continuous physical phenomena such as the behavior of elastic solids, fluids, heat, sound, electricity, and gravity.

Boolean algebra has to do with pairs of operations and includes laws for pairs of operations that make it of basic importance in computer science. The basic elements of digital computers have two states. Computer logic must therefore be built upon such things, and Boolean algebra contains the rules that are necessary to do so.

Statistics is extremely useful in dealing with large numbers of items. In manufacturing, products must be tested in order to ensure that they are within specification. One would not like to test every item, especially if the test would destroy the item (fireworks, aged champagne). Statistics allow us to determine how many we should test to have a particular level of confidence in overall quality. If one is attempting to determine whether customers will like a particular design feature, how many people and which ones should we ask? Once again, statistics will tell us.

There are many other types of mathematics (see table, page 116). Engineers use some of these, but some are little known to people outside that particular discipline. Mathematics is truly an area inhabited by specialists. According to one estimate, the American mathematical community contains 60,000 to 90,000 people, with corresponding numbers in other countries.[2] These people can submit their work to more than 1,600 respected technical journals. Stanislaw Ulam, in his autobiography *Adventures of a Mathematician*, estimated that some 200,000 new theorems are published each year in the mathematical literature.[3] Obviously no single person could keep up with this amount of new material, especially an engineer who is supposed to be doing a myriad of other things.

A great deal of mathematics is neither used nor understood by most engineers. In general, engineers have different motivations toward mathematics than mathematicians. Mathematicians are motivated by the power and beauty

A computer-generated picture of a periodic minimal surface. Such surfaces are of great interest not only to mathematicians but also to those who study the structure of materials. Modern computer graphics are invaluable in discovering, visualizing, and studying these surfaces. This one was discovered by Michael Callahan, David Hoffman, and William Meeks. The picture was generated by the Geometry, Analysis, Numerics, and Graphics Group at the University of Massachusetts, Amherst.

of the mathematics itself. Pure mathematics is one of the most aesthetic of human activities. It is a highly intellectual pursuit that is guided by the judgment of mathematicians in the particular field. Mathematicians often work as individuals and are motivated by the pleasure they get from their work, the pride from their accomplishments, and acclaim from their peers in the form of accepted publications, reactions to presentations, visibility, and perhaps even a theorem named after them. Engineers, by contrast, tend to work in groups, they tend not to spend much time or effort publishing or giving presentations for their peers outside of their own organization, and they get their satisfaction through production of successful products. There is a spectrum, of course, with applied mathematicians in the middle. People in this field are concerned with solving practical problems through complex mathematics; their services are in demand in industry, and they do publish for their peers.

Engineers use symbols, just as mathematicians do, as a shorthand way to generalize relationships in the physical world and to simplify mathematical manipulations. Once a meaningful relationship of symbols is established, numbers from many different situations can be substituted for the symbols. The formula for determining the caliber of the catapult in Chapter 1 is an early example of a relationship that allowed engineers to design machines that worked well without putting immense effort into trial and error. It is an empirical relationship, in that it is not based on any particular knowledge of nature. Experience showed it to be useful, however, and so the "formula" became a very popular tool among catapult engineers. Certainly, relations among circumference, diameter, and area and among the lengths of the sides of triangles, as well as other geometric knowledge, were necessary to early engineers. Over time, engineers have come to depend more and more on mathematically expressed relationships for their work.

Newton's second law, for example, states that a force upon a body produces an acceleration that is inversely proportional to its mass. For a given force, doubling the mass halves the acceleration. For a given mass, doubling the force doubles the acceleration. Scientists and engineers write the second law as:

$$F = ma$$

Classification of mathematics

General
History and biography

Logic and foundations
Set theory
Combinatorics, graph theory
Order, lattices, ordered algebraic
  structures
General mathematical systems

Nonassociative rings and algebras
Category theory, homological algebra

Group theory and generalizations
Topological groups, Lie groups

Functions of real variables
Measure and integration
Functions of a complex variable
Potential theory
Several complex variables and analytic
  spaces
Special functions
Ordinary differential equations
Partial differential equations
Finite differences and functional equations

Sequences, series, summability
Approximations and expansions
Fourier analysis
Abstract harmonic analysis
Integral transforms, operational calculus
Integral equations
Functional analysis
Operator theory
Calculus of variations and optimal control

Geometry
Convex sets and geometric inequalities
Differential geometry
General topology
umber theory
Algebraic number theory, field theory and
  polynomials
Commutative rings and algebras
Algebraic geometry
Linear and multilinear algebra; matrix
  theory
Associative rings and algebras
Algebraic topology
Manifolds and cell complexes
Global analysis, analysis on manifolds

Probability theory and stochastic
  processes
Statistics
Numerical analysis
Computer science
General applied mathematics

Mechanics of particles and systems
Mechanics of solids
Fluid mechanics, acoustics
Optics, electromagnetic theory

Classical thermodynamics, heat transfer
Quantum mechanics
Statistical physics, structure of matter
Relativity
Astronomy and astrophysics
Geophysics

Economics, operations research,
  programming, games
Biology and behavioral sciences
Systems, control
Information and communication, circuits,
  automata

Approximately 3400 subcategories

Source: *Mathematical Reviews*, 1979

Newton discovered this during a lifetime of thinking about the physical world and performing experiments. He did not derive it mathematically from some other relationship. Yet it tells us many useful things if properly used. If we neglect air resistance and rolling friction, the amount of force it takes to accelerate a car a given amount is directly related to its mass. An argument for lighter cars? If we have a payload in space that has a weight of 1,000 pounds on earth and includes a rocket that delivers 10 pounds thrust, we can calculate how long we must burn the rocket to give us a particular change in velocity. Let's say that we would like to speed up by 10 feet per second in order to slightly correct our trajectory and we are going slowly enough to not worry about relativistic effects.

Weight on earth is the force a mass exerts on a scale or other support under the acceleration of earth's gravity. Since we know the weight and the acceleration of gravity at the earth's surface (32.2 ft/sec^2), we know that the mass of our spacecraft is 1000/32.2 lb-sec^2/ft (the units here are sometimes called poundals—a very messy situation but the U.S. refuses to go to the metric system, which handles such things much more elegantly). Our rocket therefore gives our spacecraft an acceleration equal to the force from the rocket divided by this mass, or (10 × 32.2)/ 1000 ft/sec^2, or 0.322 ft/sec^2. Since accelerations are often described in terms of gravity at the earth's surface, this could also be called 0.01 g.

We now need one more relationship, which is that between velocity, acceleration, and time. If an object accelerates, the final velocity is equal to the initial velocity plus a quantity equal to the acceleration multiplied by the time over which it accelerates. The increment is simply the acceleration multiplied by the time. This can either be found in a book, remembered from a high school physics course, or figured out from the definition of acceleration. We are now home free. Our increment (10 ft/ sec) is equal to the acceleration (0.322 ft/sec^2) multiplied by the time. The time is therefore equal to 10/0.322, or 31 seconds.

There are a large number of simple-appearing physical relationships such as this which, when expressed symbolically, give us both insight and the ability to find quantities of interest. Ohm's law, for instance, tells us that for the flow of electricity, the voltage is equal to the current multiplied by a quantity called the resistance. This can be written:

$$E = IR$$

where E is voltage (volts); I is current (amperes); and R is resistance (ohms).

Suppose I were designing a radio to operate on an automobile battery (12 volts) and I wanted a current of 0.001 ampere (1 milliamp) in a particular circuit. I would need a resistance of 12,000 ohms from the devices in that circuit. As a more complicated exercise, I might become impatient with the time my automobile required to become warm enough on cold mornings so that the heater would do its job. What if I designed a bathroom-type heater to work in my car, hooked up to my automobile battery?

By looking at my bathroom heater, I can see that it is rated at 1000 watts. A watt is a unit of energy, and for a given device, the wattage is equal to the voltage multiplied by the current. 1000 watts at 12 volts would require a current of approximately 83 amperes. By Ohm's law, the resistance of my heater would therefore have to be 12/83, or about 0.15 ohms.

Unfortunately, if I looked also at my automobile battery I would see that its overall capacity was probably on the order of 100 ampere hours, and if my memory reached back to the time when cars were equipped with ammeters, I might remember that the rate at which batteries charge is on the order of a few amperes. I therefore might worry about installing such a heater on my automobile for fear of draining the battery more rapidly than it could be charged.

Such simple-looking laws as these, written in symbolic form, are incredibly powerful. The engineer must know when they are applicable. (For instance, Newton's law breaks down as the speed of light is approached.) Given an understanding of the assumptions underlying these laws, however, they allow great insight into the behavior of physical things. To most people such relationships look mathematical because they are written in symbols, and in fact even though they are simple physical laws, they are used by engineers as the basis for mathematical calculations.

Mathematics gives us an extraordinary language, which almost magically lets us describe very complex physical happenings of interest in technology in elegant and powerful ways—elegant because so much information is contained in so few symbols, and powerful because we can use them to derive so many relationships of use to us. Some of them are deceptively simple, such as the first law of thermodynamics, which tells us that for a given volume, the net heat entering the volume in a given time less the mechanical work done must be equal to the increase of energy within the volume. This may be written as

$$Q - W = E$$

where Q is the net heat entering the volume in a given time, W is the mechanical work done, and E is the increase in energy in the volume. But this law takes great expertise to apply properly. Supposing we draw an imaginary box around an automobile engine. If the engine is at a steady state (driving steadily along a highway), the amount of energy in the volume of the box remains constant, so that the term E is equal to zero. The first law then says that the net heat entering the volume in a given time is equal to the work done by the engine. The net heat is that which comes from the combustion of gasoline less the heat lost from the engine to the atmosphere. This tells us many things. We are no longer so cavalier about our radiator and cooling fan, as they cost us money by dissipating heat energy into the atmosphere. We prefer to travel down the road with our gasoline, not heat up the earth. The second law of thermodynamics tells us that we have to lose *some* heat, but the less the better. Knowing our gasoline consumption and the amount of energy in it, the amount of heat we lose to the atmosphere, and the work we get out of the engine, we can calculate how efficient our engine is and modify it so as to make it more efficient.

Some equations that appear simple represent impressive intellectual leaps. As an example, during the last century James Clerk Maxwell took the two phenomena of electricity and magnetism and brought them together into a coherent and unified theory. This theory describes the behavior of all electromagnetic radiation, which includes radio waves, ultraviolet and infrared radiation, visible light, X-rays, gamma rays, and other things of high interest to engineers. Maxwell's equations are essential to electrical engineering. They describe the interaction of a number of quantities: electric and magnetic fields, current density, the forces between current-carrying elements, a quantity called "electrical displacement," and time. All of these quantities except for time vary through space. However, notice how elegantly the mathematical relations are when written in vectorial form for vaccuum.

$$\text{curl } \mathbf{E} = -\frac{\partial \mathbf{B}}{\partial t}$$

$$\text{div } \mathbf{E} = \frac{1}{\epsilon_0} \rho$$

$$\text{div } \mathbf{B} = 0$$

$$\text{curl } \mathbf{H} = \frac{\partial \mathbf{D}}{\partial t} + \mathbf{j}$$

Where $\mathbf{E}$ is the electric field in vector form, $\mathbf{B}$ is the force field between current elements in vector form, $\mathbf{H}$ is the magnetic field in vector form, $\mathbf{D}$ is current displacement in vector form, $\mathbf{j}$ is current density in vector form, t is time, ρ is the charge density, ϵ_0 is the permittivity of free space.

Unless you are quite sophisticated in mathematics, you probably do not know about things like curls and divergences and equations in differential form, which these are. Unless you work with this kind of notation, you also do not have a feeling for the quantities the equations refer to. However, as you know, mathematical operations can be learned, and with enough association, abstract concepts become real.

If you were highly motivated to understand these equations, you could do so. In fact, I could probably explain the principles to you in a day or so. But the purpose of showing you these equations is neither to teach them to you nor to scare you but rather to demonstrate that an enormous understanding of phenomena of interest to engineers—in this case electromagnetism—can be absolutely described with a few symbols in a way that makes it possible to

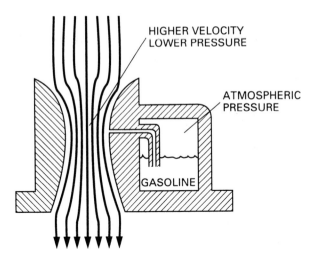

HIGHER VELOCITY
LOWER PRESSURE

ATMOSPHERIC
PRESSURE

GASOLINE

Flow in a carburetor.

use mathematics to help design equipment, understand its behavior, and think about new directions.

Unfortunately, equations such as Maxwell's are often explicitly solvable only in limited situations. The limited situations often approximate reality, however, and approximate solutions can be achieved for the general case by use of computers, which can chew away at them sufficiently to get close to answers we desire. And often if relationships are made less general, exact solutions can be found.

As an example, a set of relationships called the Navier–Stokes equations completely describe the behavior of fluid particles anywhere in a flow field at any instant of time. They are nonlinear differential equations and cannot be explicitly solved. But if we are worried about low-speed situations or liquid flow, we can ignore compressibility and the effects of viscosity. This reduces the Navier–Stokes equations to a much simpler set of equations called the Euler equations. They tell us, for instance, that for one-dimensional flow the pressure decreases as velocity increases—a perfectly good explanation for why carburetors on automobiles work and airplanes stay up in the air. Air entering an automobile engine passes through the "throat," or "venturi," of the carburetor. Because the throat is narrow, the velocity of air increases and the pressure falls below that in the bowl, which contains gasoline (see above).

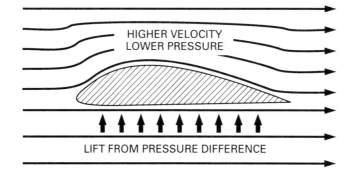

Flow over an airplane wing.

This difference in pressure forces the gasoline into the airstream. As for why airplanes stay up, the air on the top of the airplane wing travels farther relative to the wing surface in the same amount of time than that on the bottom; this means it is traveling relatively faster, and as velocity increases, pressure decreases (see figure). This differential in pressure pushes the airplane wing up, the proper direction to keep us in the air. Best of all, like other relationships we have looked at, these equations not only explain to us how things work, but they allow us to quantitatively calculate how *much* they will work.

Why does mathematics seem to mirror the physical world? Why should Maxwell's equations describe the relationship between physical electromagnetic phenomena? Why should the Fibonacci series, in which each term is the sum of the two previous terms (1, 2, 3, 5, 8, 13 . . .), correspond to the layout of pinecones and sunflowers? Why should addition and subtraction allow us to keep track of our sheep? No one knows. Philosophers have wondered about this for years, since there is no particular reason why it should. One school of thought maintains that in fact mathematical relationships are somehow built into the universe. They are "out there," and over time we discover them. Since mathematics is derived from the universe, it is no wonder that it predicts natural behavior. The other school of thought claims that we so respect our invented mathematical system that we force our experiences to fit it; we "see" mathematical relationships between phenomena that in fact may not exist in the real world. Like many philosophical arguments, there is probably some truth in each side. In any case, time and time again, mathematical relationships that have seemed to have no practical value at all have been found to be applicable in the real world of engineering. Radon's theorem, mentioned in the CT scan discussion in Chapter 2, is a good example.

During the great intellectual revolution of the sixteenth and seventeenth centuries the propensity of mathematics to predict physical reality became apparent, and mathematics became intimately linked with both science and engineering. Before then, people had taken enlightened guesses, built products, and then changed them until they were suitable. Often the results became quite sophisticated. Wooden sailing ships were so well designed that they remained virtually unchanged for 300 years. Their design evolved through a great amount of experience, and the rules were in the minds of the expert shipbuilders.[4] Common objects such as the axe and shovel are very well designed. One method sometimes used to humble design students is to assign them the task of designing better axe or shovel handles. The assignment requires production of a fully functional model. In other words, the complete project must be a functioning handle on a traditional shovel or axe blade. I have seen this done many times and look forward to presentations, since all concerned become newly impressed with the sophistication of these evolved objects.

However, trial and error is an unacceptable way to design skyscrapers, submarines, and pharmaceutical plants. For reasons of economics and safety, we must have the ability to build "in the mind" before committing to hardware. By taking advantage of mathematics, we can use basic physical principles, such as Newton's Laws, the laws of thermodynamics, Maxwell's equations, and chemical reactions, to derive relationships which not only let us understand the way different quantities interact but which allow increased understanding of what is going on.

Suppose that I offered a million dollars to the person reading this book who could most successfully tell me the smallest diameter steel wire that could support his or her weight. The hitch is that only contestants willing to hang by their wire from a helicopter hovering at 5,000 feet would be eligible. In other words, you have to be certain that the wire will hold your weight. However, to win you must pick the smallest wire that you are certain will hold your weight. Engineers refer to the ratio of strength to load as the safety factor. You want a safety factor that is small but greater than one. Is this important? Sure. Engineers in the aerospace industry often work with safety factors on the order of 1.25. What is your guess?

If you were trained as an engineer or physical scientist, your immediate reaction would be to begin calculating. But if you were not, you might begin by trying to think of examples you could use from your experience. Or you might guess. I have given this problem to many people who were not trained as engineers, and their guesses range from "the thickness of a hair" to "an inch." This is a fairly wide range for making Lear jets, or even for building decks in one's backyard.

How could we do better? Let's think about what is happening. What quantities should determine how much load a wire can hold? Certainly the material. What else? How about its cross-sectional area? We might think that the bigger it is, the more load the wire should be able to hold. That is, in fact, true. The load in the wire divided by the cross-sectional area is called the stress in the wire. Experience has also shown that this stress is surprisingly constant across the area. We might also expect that by experimenting with various materials, we could find the stress that each could take without breaking. This has been done, and the resulting values are referred to as breaking strengths, or failure stresses, of the materials. These quantities are not absolute, as different batches of the same material have some slight variation. But they are close enough to be useable by engineers.

The breaking strengths of various steels range from 60,000 pounds per square inch of area to 200,000 pounds per square inch. Let us assume that the steel wire we are using has a breaking strength of 100,000 psi (ascertained by calling the manufacturer, or testing a piece of it ourselves). We can derive a relationship between the quantities of interest by examining a little section at the top of the wire. We can apply a principle known as equilibrium. We know that in our situation, if the helicopter is holding perfectly still (we will ignore vibration for now), the little section of wire is not accelerating. In fact, it is stationary. In order for this to be the case, the forces pulling up on the wire must be the same as the forces pulling down (see figure). Otherwise there would be a net force on the piece of wire, and Mr. Newton would insist that it accelerate. We will assume that the weight of the wire is very much smaller than our weight, so we will ignore it. The equilibrium relationship tells us that $sA = W$, or $s = W/A$, or $s = W/\pi r^2$, where $s = $ stress $= $ load/area; $A = $ area; $W = $ our weight. For a breaking stress of 100,000 psi, and a load of 200 lbs (my weight), this relationship will tell me that for our example, the diameter is 0.05, or slightly under $\frac{1}{16}$ inch.

STRESS = S

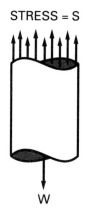

W

Segment of weightless wire.

Would I actually go out and hang by this wire? No way. First of all, any flaw in the wire or error in my assumptions would drop me from the helicopter. Second, I would worry about how steady a platform the helicopter was and how the attachments might weaken the wire. I would definitely want a safety factor. How big a safety factor would be a matter of judgment, since it would depend upon how much I value life, money, and so on. However, if I could design the attachment, if I trusted the weather and the helicopter pilot, and if the manufacturer assured me that the strength of the wire was 100,000 psi, I might be talked into hanging on a wire with a diameter of $\frac{1}{8}$ inches, since the cross-sectional area would be about four times as large.

STRESS = S

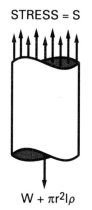

W + πr²lρ

Segment of wire including weight.

We could become fancier and learn more about our risks. For instance, were we justified in ignoring the weight of the wire? Let us call the length of the wire ℓ inches (see adjacent figure). Our relationship now becomes

$$s = (W + \pi r^2 \rho \ell)/A$$

(where ρ is the density of the wire. For steel $\rho = 0.283$ lbs/in³.) Let us assume that our wire is $\frac{1}{4}$ mile long. The diameter would have to be .052 inches. Not much change in diameter from before. We were probably all right in neglecting the weight of the wire.

If I were not hanging on the wire ($W=0$), how long would the wire have to be before it broke of its own weight? This is a different problem. See if you can figure out what the diagram would look like. The relationship would be the same as the one above, except that W would be zero. It would simply be

$$s = \rho \ell$$

The answer is about 29,000 feet.

We have seen that the simple relationship $s = W/A$, which works for any member in tension, allows us to do several things. Knowing the allowable stress and the load, we can determine the cross section of the member. Knowing the allowable stress and the cross section, we can determine the allowable load. Of course, things are not quite this simple. There are so-called stress concentrations, such as notches, holes, and other discontinuities, which cause the actual stress to be higher than the one predicted by using the cross-sectional area. Materials that are not homogeneous, such as reinforced concrete, are made of components with different acceptable stress levels.

But our formula isn't bad, and it also works for compressive failure of short members. (Long members fail by an instability called "buckling," which is a function of their geometry and their stiffness.) Some materials, such as steel, have essentially the same failure stress in tension and in compression. Others, such as wood and concrete, have very different values. Structural members fail in other ways, such as bending, shear, and torsion. Yet mathematics combined with logical reasoning and physical principles allows the derivation of relationships capable of analyzing all of them.

A final question we might ask, in considering the interface between mathematics and engineering, is what effect electronic digital computers have had on these two fields. Many pure mathematicians use computers only peripherally in their work, because they are searching for proofs and exact solutions. But other mathematicians now use them extensively and claim that if a computer algorithm can generate strong enough evidence, it should be admitted as a proof. In 1976 Appel and Haken attracted a large amount of attention by claiming to have proven what is sometimes called the four-color problem with a computer. For over a hundred years it had been conjectured that four colors were sufficient to color a map on a flat surface or a sphere so that no two countries with a common boundary would be the same color. However, no mathematician had proved it, although many had tried. Appel and Haken used a computer to examine all possible cases. The computer program and the final output were published, but the intermediate steps were simply far too numerous to ever be committed to paper. Was this an adequate proof? Mathematicians are divided. Some say yes; some say no.

Most calculation is now done by computer, particularly in the case of

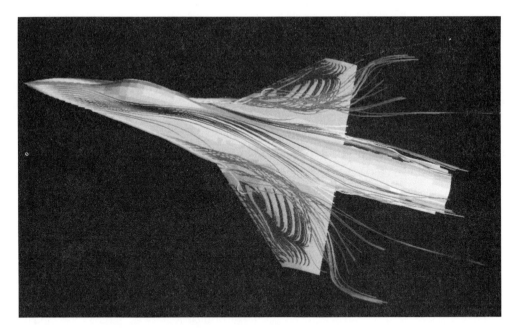

Computer-generated aeronautic shape showing particle paths.

engineering, where the philosophical rigor of an approach is less important than a useable answer. If you ask engineers what effect computers have had on mathematics, they will tell you it has been revolutionary. The computer not only has relieved the engineer of a large amount of tedious work, but has allowed much greater use of physical law and theory. The speed of the large computer makes possible iterative approaches to the solution of relationships and allows simulations of observed reality to be constructed and run either for purposes of prediction or to gain a better understanding of underlying mechanisms and their interaction. It also allows rapid and highly accurate interaction between graphical and purely analytical approaches to solving problems. The computer-generated graphic above shows the paths of particles over a possible shape for a space shuttle for a particular set of flight characteristics. The paths are based on aerodynamic theory and the calculations were done on a Cray 2 computer. Such approaches are much cheaper than testing a model in a wind tunnel, and they allow certain results not obtainable with experimentation. The computer remains dependent on humans, however. It must be fed its program and its input material, and the results are only a function of them. The computer does the calculation, but the basic concepts reside in the person.

Complicated geometries require that a number of equations be solved simultaneously. Large structural analysis computer programs, for instance, can handle on the order of 50,000 degrees of freedom, or equations. Many assumptions are necessary in order to give the computer enough information to solve the problem. The programs are typically based on the type of simple linear relationships discussed in this chapter. All parts of structures may not act in a linear way, however. It is possible to handle nonlinearities, but then one must be able to predict where they will occur and load the proper relationships into the program. Given adequate information, the computer will solve all of these equations and produce a map of loads, stresses, or other responses corresponding to the nodes. But it is only as good as the models the program depends on and the assumptions fed into the program.

The computer's ability to handle complex analysis has had an interesting effect on the design of complicated systems. For example, at the Jet Propulsion Laboratory, spacecraft were originally designed and developed with an idealized dynamic input representing the approximate behavior of the boost vehicle. Early structural analysis programs used many fewer degrees of freedom (on the order of 1,000) and therefore the input to the program was much simpler to assemble and simpler computers were able to handle the calculations. Now it is possible to analyze the spacecraft and the boost vehicle as a unit. The problem is more complex, and the cost of the analysis is therefore higher. The analysis also considers the interaction of many more components. The result is undeniably a much more accurate analysis, but some engineers claim that it is now extremely difficult to make changes in their subsystem during development because of the necessity of re-doing a very expensive procedure and modifying an increased number of other subsystems. Therefore, while analytical sophistication, and the optimization of a particular system design, is undoubtedly increasing, flexibility in design modification may be decreasing. As systems become more optimum, the individual engineer must be more careful.

One Student's Answer to the Problem on Page 109.

A toilet tank looks as though it contains about five gallons of water. Probably, on the average, people flush a toilet about five times a day and there are 250,000,000 people in the country. This requires $5 \times 5 \times 250,000,000$, or $6\frac{1}{4}$ billion gallons a day. This is significant.

6

Science, research, and engineering are part of a continuum. The understanding gained by science often serves as a basis for technological development; and yet modern science could not operate without technology. Research, which implies a careful, studious pursuit of better understanding, is found in both science and engineering (as well as in most other fields of endeavor). Engineering research and scientific research often overlap. Indeed, the familiar term "research and development" (R&D) can cover the entire gamut from pure science to product development.

The relationship between science and engineering is more subtle than that between mathematics and engineering. It is perhaps easier to discuss, since science does not seem to evoke as strong negative emotions as mathematics. One hears much more about "math phobia" than about "science phobia." Yet it is not clear that most people understand the nature of science or its relation to technology. Many seem content merely to blur the two terms or obscure their relationship with such statements as "technology is the application of science." These misconceptions annoy people like me, who as an engineer wants to be thought of as more than someone who hangs around waiting for science to produce something brilliant so that I can apply it. It would be nice if science and engineering could be treated as parts of a continuum. Unfortunately, they are separated in educational institutions, in industry, in the minds of government legislators who make decisions concerning funding and national priorities, and in the thinking of much of the public. Because of this, it is worthwhile examining their interaction in more detail.

The relationship between science and engineering varies with the product. Little or no science is involved in many technological activities (house construction or the making of nails, paperclips, and automobile hubcaps). Quite a bit of science goes into the production of plastics and the design of water treatment plants. In areas such as genetic engineering, science leads technology, although the interaction between the two is subtle and complex. The relationship between science and engineering tells us quite a bit about the nature of present-day technology.

Part of the confusion results because people use the word "science" to describe a number of different things. There is first of all, scientific knowledge. This takes the form of scientific laws, theories, and hypotheses. A

The main beam tube of the Stanford Linear Accelerator, a powerful scientific instrument based on sophisticated technology. This device has the capability of simultaneously accelerating both electrons and positrons to an energy of 50 billion volts (a flashlight accelerates electrons to a few electron volts) to allow collisions involving 100 billion electron volts. The accelerator and associated laboratories cost $114,000,000 when constructed in 1966, have been upgraded and modified several times, and have contributed to two Nobel-prize-winning discoveries in physics.

scientific law is a relationship that can be verified by experiment and that allows prediction. Newton's laws of motion are a good example. In the domain in which we live (velocities much less than the speed of light, objects larger than quarks and smaller than galaxies), Newton's laws predict physical motion and its relation to forces. Laws are not perfect, in that we usually have to make simplifying assumptions about reality for them to work. Even so, they are dependable enough to be used quite successfully by engineers.

Scientific theories are a bit more tenuous. Like laws, they evolve from attempts to make sense out of careful observations, but for various reasons, verification through experiment is extremely difficult or impossible, and therefore theories do not have quite the same status as physical laws. For example, Robert Boyle (1627–1691) found that for a given temperature, pressure in a gas is proportional to its density, or inversely proportional to its volume.[1] Later, J. A. C. Charles (1746–1823) discovered the complementary relationship—that for a given volume, pressure increases as the temperature increases. These are now called Boyle's and Charles' laws and can be "proven" by experimentation in a very straightforward way. They are certainly useful to an engineer attempting to figure out why a boiler has exploded or how to get more power out of an engine, but why are they true?

In 1738 Daniel Bernoulli realized that if a gas consisted of molecules in motion, pressure would be generated as these molecules banged against the walls of their container. This kinetic theory of gases was improved by James Joule (1818–1889) and R. J. E. Clausius (1822–1888) and is a convincing explanation for Boyle's and Charles' laws (more molecules hit a given area at a higher density, and each hits harder at a higher temperature). At the time this theory was developed, it was strictly a mental construct, since there was no way to count or clock molecules. At present, it is consistent with so many other scientific theories that we accept it almost as fact. Yet it is still a theory, and we would be less surprised to find it superseded by a new theory than to find the pressure of a fixed volume of gas falling as the temperature was increased.

Theories seem reasonable in the context of other things that we know to be true or strongly believe, but they have not been, or cannot be, proven by experiment. They are generally supported by a large number of scientists familiar with the evidence in great detail. Some fields of science, such as cosmology, geology, astronomy, and evolutionary biology, are based on theories, since experimentation is essentially impossible. Hypotheses have not yet attained the status of theory. They are still guesses or ideas and have not yet been subjected to rigorous analysis through either long-term observation or experiment. Hypotheses rarely appear in introductory textbooks about science, whereas laws and theories have starring roles.

Despite their unproven status, some theories and hypotheses are of great use to engineers. Materials science, for instance, includes many theories that allow the properties of materials to be varied over wide and useful ranges. Other important scientific theories, such as the big bang theory of the origin of the universe or Darwin's theory of evolution through natural selection, are intellectually fascinating to scientists and engineers alike, but not of much practical use to engineers in their daily work.

The word "science" is also used to refer to the scientific process, or the approach to problem solving used in science. The exact nature of this process is the subject of much debate among philosophers (as is the nature of scientific knowledge), but it is certainly the case that scientists do think in a particular way. Among other definitions, science has been called the search for consensus, and scientists therefore like to measure and work with quantity since quantitative information is easier to compare and contrast than qualitative information. Scientists produce ideas about the nature of things and attempt both to justify their own ideas and disprove the ideas of other scientists. They also rely heavily on the language of mathematics and have a certain sense of ethics about the use of scientific information (one does not falsify data, for example). Much of this scientific process is similar to the problem solving done by engineers, which is not surprising since engineering education places heavy emphasis on science. Because of the overlap both on the knowledge side and on the process side, it is no wonder that simple attempts to relate science and engineering get into trouble. Let us refer to a bit of history again and think about the relationship of the two over the ages.

Humans have probably observed their world and tried to make sense out of it for most of their history. However, their interests were pragmatic—when to change hunting grounds, when to plant crops, when to have ceremonies to ensure proper characteristics for children—and their explanations of the natural world were based on supernatural beliefs until fairly recently. In the West, the Greek natural philosophers were the first people we know of who approached the world in a manner akin to modern science.

But they were, first and foremost, philosophers, in that they were interested in the purely mental aspects of the attempt to understand the natural world. Plato, perhaps the most influential of them all, believed in a level of theoretical perfection that could not often be seen in the outside world. His thinking affected natural philosophy for a very long time, and is still alive and well in the faith that scientists have in beauty and simplicity. The early Greek philosophers also were somewhat suspicious of hypotheses. They were in love with logic, and thought that knowledge of nature could be logically determined from universally accepted "facts." Aristotle was the most powerful spokesman for this approach to natural philosophy and spoke with such force that his ideas lasted for 2,000 years. He regarded deduction as the highest form of logic. The Greek philosophers were fascinated with deduction because if the premises are correct ("All men are mortal. Socrates is a man.") it produces a correct answer ("Socrates is mortal."). It should be possible, they thought, to deduce true natural laws from similarly undebatable premises, a flattering if unattainable ideal for scientists.

Many people have subscribed to this faith in deduction over the ages and tend to treat science as an activity that establishes eternal truths through the application of impeccable logic to undisputed facts. Some of my best students are surprised when they realize that the "truths" of science are constantly being improved and made obsolete. This attitude is not surprising. For some years science has been an article of faith of our society, and scientists have been given credit for such things as decoding the genetic program of life, explaining the structure and origin of the universe, and discovering the most fundamental particles of matter and energy. Science teachers like to have their students deduce relationships from "basics," which imply that the building blocks are all known. Unfortunately, that's not the case. We don't always know how to make the proper logical argument to get from a premise to the desired conclusion. Worse yet, we don't have any idea what all the premises are; we aren't even close to knowing.

When deduction would not work, the Greeks considered induction to be the other acceptable form of reaching understanding. Induction is a somewhat weaker form of argument, in that it is not illogical to accept the truth of a premise ("All crows I have seen so far have been black") and yet deny the truth of the conclusion ("All crows are black"). The strength of the conclusion depends on the strength of the premises—on the size of the sample, the care of the observations, and so on—but it can never deliver results that are completely ironclad. Even here, though, we get into trouble, for over the history of science, major "correct" premises ("The earth is the center of the solar system." "The four basic elements are earth, air, fire, and water.") have repeatedly proven to be wrong, and, as we said before, we don't have enough of them anyway. Science demands hypotheses derived from attempts to explain inconsistencies—guesses, if you will. Induction also requires observational or experimental validation as well as mental argument before it can work in science. Deduction and induction are not enough. Scientists must actively deal with conjecture, uncertainty, randomness, observation, and experiment.

It was not until the time of Francis Bacon (1561–1626) that the adequacy of the Aristotelian approach to science was questioned. Bacon was one of the earliest true philosophers of science. Although, like Aristotle, he thought that science should progress through use of logic to derive relationships from agreed-upon observations, he also believed strongly in the importance of experiment to check scientific conclusions. Bacon lived at the beginning of what is called the Scientific Revolution, during which a quantitative, mechanistic conception of the universe displaced the divinely ordained universe of Aristotle and the more mystical concepts of Plato. Bacon therefore was too early to witness the progress in understanding that came with the conceptual ability of Galileo, Boyle, Newton, and others. However, he definitely was seeing the directions that scientific investigation was to take.

During the Scientific Revolution, science for the first time began to overlap technology. The knowledge that was produced by people such as Newton and Boyle was of use in technology, and the processes used by scientists and those involved in technology began to converge. Scientists got their hands dirtier, and early engineers became more methodical in their attempts to understand phenomena. Both relied more and more on conceptualization, logical argument, and experiment.

For example, in the development of the steam engine, discussed briefly in Chapter 4, Savery's pump and Newcomen's engine no doubt owed a debt to previous work done on vacuum and air pumps by people such as Toricelli, von Guericke, and Huygens, who are considered to have been scientists. Yet much of their work was in response to technological phenomena that mystified people (such as the failure of "vacuum" pumps to lift water over 32 feet). Since Newcomen was an iron monger and his partner Calley was a plumber, their machine was a mixture of the products of early science and mechanical ingenuity.

James Watt realized that a great amount of energy was lost in the Newcomen engine because the water that was introduced into the cylinder to condense the steam also cooled the cylinder, which then had to be reheated by the next charge of steam. He approached this problem in a rather scientific way, by considering a number of alternatives, deciding upon a separate condensor, and then spending many years experimenting with his concept until he understood it well enough to implement it. Whereas Newcomen developed his engine in six years (1709–1715), Watt spent seventeen years (1767–1784) working on his externally condensing engine. During this time he also made the engine double-acting (steam alternately on each side of the piston) and expansive working (stopping the flow of steam into the cylinder early in the stroke), both of which increased the efficiency of his engine. Watt made much better use of the "scientific method" than Newcomen. Maybe the reason he is given so much credit for development of the steam engine is that he thought much more like modern engineers than did most of his contemporaries. The next step in steam engines was to use high pressure. Such an engine was first built by Richard Trevithick in 1798, followed by a successful locomotive engine the next year. However, he was somewhat of a throw-back to earlier approaches to developing machinery. He built it but did not attempt to understand the physical phenomena in detail.

Sadi Carnot, the founder of thermodynamics and a true scientist, was to study the high-pressure steam engine and finally understand its workings. He devised the Carnot cycle (see figure), a theoretical change of state that defines the limit of efficiency in a heat engine, and he formulated the concept of thermodynamic efficiency. The vertical axis of the figure represents pressure in a gas, which for our purposes we will assume is in a cylinder. The horizontal axis represents the volume in the cylinder. Carnot's cycle is an idealization. During the first part of the cycle (line 1-2) the gas is brought into contact with a heat source that stays at constant temperature while the gas expands, absorbing heat and producing work by pushing on the piston. During the next portion of the cycle (line 2-3), the cylinder is insulated and the gas continues to expand and do work without gaining or losing heat. During the third portion (line 3-4), the gas is brought into contact with a cold reservoir that stays at constant temperature while the gas contracts, carrying the piston inward and losing heat to the cold reservoir. During the final portion (line 4-1), the cylinder is again insulated and the piston continues to move inward with no gain or loss of heat. The area inside of these lines (shown shaded in the figure) represents the net work that is done by the piston during the cycle.

This cycle demands perfect processes that cannot be obtained by real engines. Heat sources and reservoirs do not stay at constant temperature and perfect insulation is not obtainable. In addition, energy is lost through friction. However, it gives us tremendous insight into the nature of so-called heat engines. Carnot's cycle is the best that one can do with regard to efficiency. It represents an ideal. Engineers use it both as a target and as a comparison to optimize real engines. It tells us some simple truths about so-called heat engines. The efficiency of the Carnot cycle is the work done divided by the heat required. This can be derived from the first law of thermodynamics and some simple relationships that describe the behavior of gases. The result is

$$W/Q = 1 - T_c/T_h$$

where W is the work done per cycle, Q is the net heat into the gas, T_c is the cold reservoir absolute temperature, T_h is the heat source absolute temperature. If we look at a modern central power plant, we find that the hot temperature is limited by the machinery to something on the order of 800 degrees Fahrenheit, which corresponds to 1,260 degrees Kelvin, an absolute temperature scale (zero on this scale is at absolute zero, a theoretical temperature at which all molecular motion stops). Since these plants typically use

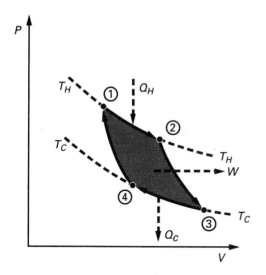

The Carnot cycle.

an ocean or river as a cold reservoir, T_c is approximately 70 degrees Fahrenheit, or 530 degrees Kelvin. The Carnot efficiency, which is the best obtainable for a heat cycle, would therefore be about 60 percent. This means that a plant such as this will put at least 40 percent of the energy in the fuel into the ocean or river. In actuality, efficiencies of these plants do not exceed 40 percent. Incidentally, thermodynamics also speaks to a common question asked by some nontechnical people: Why dump the heat in the ocean instead of making it into electricity? The answer is that you have to. It is impossible to run a heat engine without a cold reservoir, and since central power plants must dump a large amount of heat into such a reservoir, they tend to be located on oceans and rivers.

The science of thermodynamics was born over a hundred years after the steam engine. It produced understanding that allowed engineers to develop steam power to the level of efficiency and scale seen in the modern power plant or ship propulsion system. Modern engineers who develop power plants have acquired this scientific knowledge and use the scientific method to extend it. But scientists have in the meantime become much more sophisticated in dealing with basic issues such as the behavior of individual particles. Engineers often avoid having to consider microscopic behavior by applying macroscopic approaches. For instance, they often can reach acceptable answers by considering what enters and leaves a particular volume in space (called a "black box"), rather than understanding exactly what is happening inside. A physicist, by contrast, would want to know what was going on in there.

The difference in the approaches of scientists and engineers can be seen clearly in the fascinating concept of entropy. Entropy is a measure of disorder. The entropy of any system can be described by a number, and the higher the number the less energy is available for doing work. In an ideal process in which there are no losses due to friction, nonequilibrium heat transfer, and such things, the quantitative change of that number with the addition of a tiny amount of heat is dQ/T, where dQ is the amount of heat added and T is the absolute temperature at the time. In the figure on page 137, lines 2-3 and 4-1 are isentropic (constant-entropy) processes. The increase in entropy in the process described by line 1-2 is equal to the decrease in that described by line 3-4, so that the cycle returns to the same value of entropy at which it begins. This is another way of describing a perfect, lossless, ideal cycle—the entropy does not increase. In a real system, the entropy does increase because of friction, nonequilibrium heat transfer, and other losses. If I have an isolated real system containing a hot reservoir, a cold reservoir, and a heat engine doing some sort of work, both reservoirs will eventually reach the same temperature and the engine will cease running. The entropy of the system has increased, and energy is no longer available to do work.

The second law of thermodynamics states that in an isolated system the entropy change must always be greater than or equal to zero (greater for a real system in nature). It tells us about the impossibility of such things as perpetual motion. If I drop my shoe on the floor, the first law of thermodynamics will tell me how much warmer the shoe and the floor will become through absorbing the kinetic energy in the shoe. However, it will not tell me why the shoe cannot bound up in the air again as the temperatures return to their pre-impact state. The second law does. It also tells me that my power plant must dump heat into the ocean or river. The second law and the concept of entropy are used extensively in the design and development of heat engines, and tables of entropy values have been compiled for different states of fluids commonly used in such applications.

Physicists think about entropy in a different way. They define entropy as the measure of the relative probability of a particular state occurring. As a system evolves from a less probable configuration to a more probable one, the increase in probability is equal to the increase in the entropy. This view of entropy is microscopic, and has to do with the probable behavior of particles. Quantitatively, the entropy, S, is proportional to the natural logarithm of the probability, and the proportionality factor is a number called Boltzmann's constant, a relationship that is engraved on Boltzmann's gravestone: $S = k\ln P$.

Physicists and engineers agree on, say, the entropy of a perfect gas in a particular state. However, the scientist looks at the statistical behavior of the particles, and the engineer looks at the behavior of the whole. A physicist would say that the likelihood is overwhelming that heat flows from a hot object to a cold object, but in principle there might be a tiny possibility that the opposite might happen. The possibility is so small, due to the large number of particles participating in real processes, that they would certainly be surprised if it did occur. Engineers tend to say that the possibility of such a thing occurring is zero.

Incidentally, attempts to extrapolate the concept of entropy beyond thermodynamics and statistical mechanics are quite common. Students sometimes believe that their rooms constantly become messier because entropy must increase. I have seen the concept of entropy used to explain why large organizations become mired in increasing bureaucracy. Such applications are usually made by people who do not really understand the concept, especially the importance of defining the system carefully. But they are a good example of the mystical effectiveness in our society of esoteric scientific and technical concepts, even if used improperly.

Most engineers do not have the understanding of physicists at the microscopic level. They are, if you will, behind the scientists. Yet most scientists would have trouble producing an engine without being distracted by their curiosity to understand basic phenomena better. Once again, in the development of modern engines, it is difficult to draw the line between engineering

Engineers in a "clean room" manipulate the primary mirror of the Hubble Space Telescope during production. Ground tests indicated that the mirror's reflectivity exceeded specifications. If the Pacific Ocean were as smooth as the reflective coating of the ST mirror, the largest wave would be only a thousandth of an inch high. And yet after launch, it was discovered that the mirror had been ground to the wrong contour. The performance of this highly sophisticated and expensive scientific instrument has so far been handicapped by an error made in its design—a topic to be discussed in Chapter 7.

and science. Most scientists work for a level of understanding that most engineers do not need, and most engineers do productive work with very little knowledge of the basic phenomena. If you want a quick insight into how productive a nonscientist can be, find someone who modifies engines for drag racing. These mechanics accomplish technological miracles without having ever heard of Sadi Carnot.

Returning to our chronological discussion of the nature of science, in the early nineteenth century, William Whewell (1794–1866) proposed a model for scientific problem solving that is known as the hypothetico-deductive method. This model recognized that science depends upon hypotheses, rather than undebatable facts, which then are logically argued and verified or disproved through experiment. Philosophy was discovering modern science. There have been many attempts by philosophers to define the scientific process since Whewell. They have generally become more pragmatic. A man who was extremely influential in the attempt to understand science was Karl Popper. He believed that science was a process of testing hypotheses. The more ways hypotheses could be tested the better they were, but they could never be proven. Science was a search for falsifications. The more falsifications that were uncovered, the healthier was science. His image was of scientists observing, creating hypotheses, and testing them until they eventually failed.

Thomas Kuhn, in his classic book *The Structure of Scientific Revolutions,*[2] disagreed with Popper and came up with an even more worldly explanation. He claimed that science moves in new directions because established theories have increasing difficulty dealing with new observations over time. Yet new theories that explain away these difficulties are resisted by most scientists, who prefer to refine and extend prevailing theories, not to prove them false. "Normal" scientists have sufficient vested interest in the prevailing theories to argue against sweeping new directions. Certainly much evidence demonstrates that most scientists do act in this way. A good example of this was the resistance of earth scientists for many years to the theory of plate tectonics (shifting continents). Only after the evidence became overwhelming was it accepted, and it is now doctrine.

An even more extreme view of science than Kuhn's is that taken by Paul Feyerabend, who has referred to himself as a Dada-ist. His philosophy is sometimes described as "anything goes." He claims that science is not rational; that philosophers like to find patterns in it, but scientists are oblivious to such grand logic. Certainly one can find a large component of irrational or illogical behavior in science, and an even larger one in engineering. Though broad principles of experimentation and recordkeeping are generally adhered to, and though most scientists could identify "good" science and "bad" science and the criteria upon which they make the judgment, practicing scientists do not brood on the scientific process as philosophers do. They simply do it, as do engineers.

In a book entitled *Big Science, Little Science,* Derek de Solla Price first showed the exponential growth of science.[3] He looked at indicators such as number of scientists, resources dedicated to science, number of papers and journals, and so on, and concluded that the growth was much faster than is usual for human activities. As an example, for every single paper published in 1665, there were on the order of 100 in 1765, 10,000 in 1865, and 1,000,000 in 1965. This is a factor of 100 increase every century. We quantitative thinkers realize that this explosive growth cannot continue forever, and in fact it is already tapering off. However, the size, the visibility, the effect, and in fact the nature of science has changed dramatically in a short time, as is true of technology. Science before this century was dominated by giant individuals—Darwin, Lavoisier, Boyle, Volta. Science is now dominated by huge projects and laboratories with high levels of government, foundation, or private funding—the proposed supercollider and the human genome project are two highly visible examples of this. In fields such as physics and medicine, the single-author paper is a vanishing breed. Publications in experimental physics may list dozens of authors, a fair testimony to the impossibility of separating personal contributions from the common goal. Like technology, science is conducted by large organizations that are continually engaged in obtaining resources and with reporting findings fast enough to please the funding agencies. In *Philosophy and Sociology of Science,* Stewart Richards estimates that no more than 10 percent of all scientists are still working individually in the traditional manner.[4]

Another change has been the increasing amount of science that is done in industry. The last half of the nineteenth century saw the establishment of large company research organizations, first in Germany and then in the United States. Now we take institutions such as Bell Laboratories and the huge research efforts in companies like IBM, Exxon, and Monsanto for granted. In industry, the line between science and technology often disappears. Doing basic research are pure scientists, whose allegiance is to their field, to their participation in the reward system of publication and peer acclaim, and to the long-term implications of their work. Technologists also exist in the pure form in industry. Their allegiance is to production of economic products, and they lack interest in publication and peer recognition. They do development. In between, basic researchers work in areas that are contiguous with company plans, while applied researchers conduct scientific investigations oriented toward a particular product line.

This distinction can be clearly seen in industries based on biomedical engineering. For example, investigation of the amino acid sequence in an antibody molecule would be called basic research. The use of this information to distinguish between the antibodies that respond to various diseases would be applied research. The finding of a technique for synthesizing the antibodies for particular diseases and testing their effectiveness would be called development. This would be followed by design and production of the packaged drugs to fight the diseases.[5]

Research and development is often treated as a separate definable activity from other aspects of a company's work, such as design, production, and operations. In general, R&D is an activity that contributes to future directions more than present products; it is managed differently because it is subjected to different expectations. Separating R&D also gives an organization an easy way to monitor the percentage of effort spent on learning to do new things: on observing, experimenting, theorizing, and publishing papers.

Engineering schools recognize the overlap in industry between engineering and science, and they design their curricula accordingly. Engineering education is strongly theoretical and geared toward mathematics and science. This is partly because of the natural interests of people who are attracted to a professorial life and who set the curriculum. It is also because engineers can learn the more applied portions of their field on the job, while they are unlikely to learn math and science on the job. But because the activities of the engineering student have little relation to the activities of many practicing engineers, it is likely that engineering education discourages some students who would make excellent engineers and encourages some others who will not. The mentality to do well in engineering schools emphasizes the ability to work problem sets and get right answers. In engineering, there are never right answers and few problem sets.

Let us finish the chapter by focusing a bit more on integrated circuits and genetic engineering—two recent developments that have very much affected our thinking about science and engineering. In both cases, stunning new directions in technology and entire new industries have grown out of new scientific insights, in one case insight about the nature of materials and in the other about the nature of living systems.

The invention of the transistor, in 1947, was based on new scientific understanding of the behavior of electrons in materials, specifically semiconductors. Yet the transistor work also required technology: the ability to manufacture carefully controlled semiconductor material and extremely precise power supplies and measuring instruments. Since 1947 we have seen tremendous refinement in the properties of semiconductor devices. The integrated circuit, invented in 1958, contains many devices, both active (transistor) and passive (resistor), on a single chip. The advantages of this integration are (1) decreased cost, since the devices can be manufactured simultaneously in a single sequence of processes, (2) decreased size, which allows lower power use and faster response speed, and (3) increased ruggedness and reliability, due to smaller physical masses and greater process control.

Since the invention of the integrated circuit, the number of devices per chip has been growing exponentially, as the size of each device shrinks exponentially.[6] At present some integrated circuits contain on the order of one million devices. This change is slowing down because of the problems associated with manufacturing devices with dimensions as small as one millionth of a meter. But the changes this has caused in the electronics industry have been phenomenal. The cost per bit in read-only memory has dropped from approximately one cent in 1970 to approximately one one-thousandth of a cent now.[7]

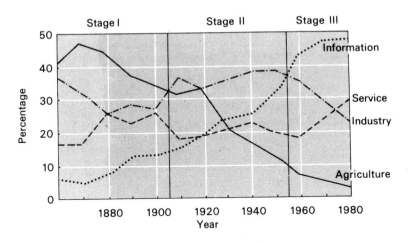

Changes in the work force from 1880 to 1980.

Factory sales of integrated circuits has increased from $10 million in 1963 to $10 billion at present, and factory sales of electronics in general has increased from somewhat over $10 billion to over $200 billion in the same period.[8] This development in technology has been essential to the so-called information age, which is having major effects on our lives. The relatively recent increases in both the information and service sectors, as well as the decreases in the industry and agriculture sectors (see figure), are all resulting partly from continued developments in integrated circuit technology.[9]

Our ability to investigate genetic material has also changed remarkably since 1953, when James Watson and Francis Crick discovered the structure of DNA.[10] Technology was also involved in their discovery. Although they were able to construct their structural model from low-tech materials such as wire and simple bits of metal, the arrangement depended heavily on the X-ray diffractometry data of Maurice Wilkins and Rosalind Franklin. In the early 1970s scientists learned how to isolate and recombine specific segments of DNA, and this opened up tremendous commercial potential. It would be possible to change living organisms such as bacteria so that they could better produce products desired by humans, such as insulin and human growth hormone. In addition, it would be possible to change humans directly. The potential of this was so overwhelming that a conference was held at Asilomar, California, in February 1975 to discuss the risks of this research. Two years later, the National Academy of Sciences hosted another meeting, also accompanied by controversy, on the techniques of genetic engineering, or the application of recombinant DNA (gene splicing). Today, the concerns still exist, but we have embarked on considerable application of these techniques.

Over 200 companies now participate in genetic engineering in the United States. A tremendous amount of knowledge and sophistication has been gained in the last two decades. Genetically modified microorganisms are being used to produce interferons, neuropeptides, blood clotting and antishock materials, vaccines, antibiotics, and other weapons against disease. Major applications in the food, basic chemical, and energy industries are expected. Pest-resistant vegetables are under development, and our future fuels may be produced by genetically altered bacteria happily munching genetically altered plants.

The knowledge gained in this period has also led to the production and use of proteins known as monoclonal antibodies, which are produced not by recombinant DNA techniques but rather through the fusion of a tumor cell with an antibody-producing white blood cell. These clones live virtually forever and produce chemically identical antibodies, which are used in a wide range of research and in diagnostic tests and will undoubtedly gain more application in treatment.

The most controversial application of recombinant DNA technology is in the treatment of human diseases. Already physicians have the capacity to change the DNA of defective bone marrow, so that it produces adequate numbers of healthy white blood cells; and there are many other diseases that potentially can be avoided or cured through direct application of genetic engineering. But genetic engineering also raises the possibility of more profoundly changing the characteristics of individuals and even of changing the germ line, so that future generations would be affected. The philosophical and ethical issues are tremendous and are resulting in ongoing discussion and regulation. Yet the potential for good is tremendous.

As in the case of integrated circuits, there is a symbiosis of science and technology here. The science feeds the industry, and technology feeds the

Scientists or engineers?

science. The photograph above was taken in a laboratory in the Beckman
Center at Stanford University. This is a central facility to support biochemistry
research in various portions of the medical school. Are the people in the
photograph scientists or engineers? Probably they are scientists, since the
majority of research in the facility would fall under the rubric of science.
However, notice the instruments. These are highly sophisticated technological
devices that drastically increase the speed and accuracy of the research worker.
The laboratory makes use of protein sequencers, for instance, which are
instruments that immobilize a protein and then remove and analyze one amino
acid at a time. Proteins are extraordinarily complex molecules, and these
instruments can describe their complete structure. They are capable of performing an analysis with fractions of a microgram of protein. The laboratory
also uses instruments to measure the purity of samples (capable of analyzing

a quantity as small as one billionth of a gram), find the relative amounts of each amino acid in a sample, and synthesize peptides. The synthesizer can build molecules of protein up to 100 amino acids long. The experimenter simply types in the desired structure, and the synthesizer then assembles the molecules in batches of useable size (hundreds of milligrams).

The laboratory also has instruments to analyze and synthesize DNA. DNA is an extremely long molecule made up of the bases thymine, adenine, cytosine, and guanine, arranged in a highly variable sequence. The analyzer can take up to 10 different DNA samples and analyze up to 500 bases in a sequence in a day. This leads to a description of the type and location of 5,000 bases in eight hours at very high sensitivity. The synthesizer can build 4 different batches of DNA material at once, each up to 70 bases long, operating at a speed of 6 minutes per base, which is extraordinarily fast compared with the time it would take to do the work by hand.

Another extraordinary bit of technology in the laboratory is a fluorescence activated cell sorter, or FACS. The operation of this machine requires that a liquid pass through a tiny orifice in a vibrating chamber, so designed that the liquid emerges in droplets, each one containing a cell or not. The droplets are then illuminated with a laser, and an array of detectors senses information about the size and distribution of the cells. By tagging the cells with fluorescent antibodies, it is possible to identify and compute quantities of various cells and, by using charged plates, to sort them into various containers. All of these instruments make heavy use of computers and automatically perform the necessary chemical procedures. All of them allow scientists to learn things about the nature of life that they otherwise could not and to proceed at drastically increased rates. Biochemistry is an excellent example of a science that is greatly aided by technology.

Developments such as the integrated circuit and genetic engineering, coming on the heels of other relatively recent developments such as nuclear engineering and space exploration, have caused us to think of science and engineering as intertwined. As I noted previously, that is a reasonable perception, but educational institutions, funding agencies, personnel departments, and in fact much of society still cling to the notion that they are separate. But the only meaningful separation that can be done is on the basis of goals, expectations, professional structure, and reward system. Scientists seek knowledge for its own sake, although they are affected by practical considerations ranging from funding source to the reactions of the media. Engineers seek to produce marketable products (including sophisticated products used by scientists), although they frequently need to understand things better in order to do so.

Let us now consider development, testing, and failure in the context of engineering. Experimentation (and failure) are common to both science and engineering. But experimentation generally has a different motivation and focus in engineering. The next chapter will examine this in more detail.

7

Development, Test, and Failure

The Proof of the Pudding

Scientists do not have a monopoly on experiment. Technology also depends on it. Most of this takes place in planned testing, but not all. Technology is as much an art as a science, and despite the rather impressive ability of engineers to predict by use of past experience, theory, and computer calculation, reality has a bad habit of not acting quite like our theories and simulations. A certain amount of failure is, and always has been, part of the game. In order to minimize failure, engineers would have to attempt nothing new, and even then things would break and people would screw up.

In this chapter we examine experiment, test, and failure at several levels, beginning with the routine sorts of things that engineers do and ending with the large-scale tragic failures that capture the attention of the media and the public. The message in this chapter is an important one, because while no one likes failures, they are an unavoidable part of technology. Being human, our society would like the benefits of technology without the failures: ever cheaper air transportation with no crashes, more sophisticated weapons with no overruns, increasingly abundant energy with no pollution, and so on. The public has great faith in technology, perhaps greater than is consistent with the nature of engineering. To prevent failures completely can be impractically expensive or even impossible.

So what kind of experiments do engineers perform? Engineers may experiment simply to discover a solution. One person who became famous for this approach to problem solving was Thomas Edison, who in his search for a practical filament for his light bulb tested a very large number of materials. Let's suppose that you, as an engineer, might need a coating to withstand a peculiar environment, and find that you can best solve the problem by trying a bunch of them in the environment. You may have done this with paint for a portion of your house. All of the vendors claim that their paint is the best. The only way one can find the truth is to try several of them. Engineers do a lot of this kind of trial. For instance, one may have several attractive solutions for a product's packaging and find that the best way to choose is to make a prototype of each of them and solicit customer reaction. If one has no pertinent experience to draw on, and one cannot find an applicable theory, discovery is about the only option left. There is no discovery without trial, and when trial occurs, so may error.

On the afternoon of October 22, 1895, the express train from Granville (in Normandy) could not stop as it arrived at the Gare Montparnasse in Paris. The train went into the waiting room, and the locomotive, emerging through the station's exterior wall, fell into the Place de Rennes. A newspaper seller on the sidewalk of the Place de Rennes was crushed to death.

It is not unusual for engineers to find themselves with no decent theory and no experience with a similar situation. The engineer attempting to predict the lifetime of a new product, for instance, is in such a situation. It is rather embarrassing that after all of these years, we have no good theoretical way to predict something so critical to product development as the phenomenon of wear. If you were to begin rubbing your knife and fork together at dinner tonight, no engineer could tell you how fast they would wear, even if you carefully controlled factors such as the cleanliness of the surface, the pressure between them, the relative velocity, and the geometry. The only way to find out would be to test.

Remember Dennis Boyle's spring and friction mechanism which held the top of his computer in any position in Chapter 5? Even though the torsion from the spring was reasonably calculable, the friction device was a different story. The theory simply was not adequate to allow the drag device to be sized closely enough. It was therefore necessary to experiment to find the proper dimensions. Some of the configurations built and tested are shown in the figure. After a choice was made, a mock-up was built and the assembly tested through 5,000 cycles. During these tests it became apparent that the friction was changing due to galling of the metal surfaces. This required introducing a lubricant, which of course changed the friction and required the process to be done all over again.

Even where theory exists, often no one in the group is familiar with it, or it may be uneconomic to obtain the necessary knowledge. Engineering is a pragmatic activity, and there is never the time, energy, or expertise to ensure that all approaches are being considered. When I attended engineering graduate school I was amazed to find that theory existed that would have been invaluable in the work I had been doing in industry before that time. But since I hadn't known the theory existed, I had simply solved my problems as best I could by judgment, trial, and error (experiment) in blissful ignorance, because that was my job. Had I known the theory existed and asked my boss to send me to school to learn it, he would have told me that we didn't have the time (much less the money). By the time I received my Ph.D. degree, I had a better idea of the scope of engineering knowledge that was available. However, I had also accepted the fact that not only would I never know more

Prototype springs.

than a fraction of this knowledge, but I would never know that much of it even existed. Therefore, I returned to being comfortable living in a state of blissful ignorance, although perhaps a more enlightened state. I do not want to even think about the amount of time I have spent in determining things experimentally that the right person could have predicted theoretically.

Experimentation is also vital in developing new theory. Research in engineering is typically an integrated attack using both analysis and test. Some theory is based entirely on experiment. I did my Ph.D dissertation on remote control, a topic of great interest at the time because of the desire to remotely control machines such as roving exploratory vehicles on the moon and other planets. There is an unavoidable time delay between the time a signal is sent to a machine in space and the time it is received, and an equivalent delay before feedback is received. The roundtrip delay for the moon is on the order of three seconds. At the time many plans were blithely being made to control such vehicles by commands from a human operator using television as feedback. I suspected there might be difficulties and did my research on the problem. I found lots of difficulties. In fact, given the proper combination of vehicle dynamics, time delay, and speed, people simply cannot control such things directly. A tremendous amount of help is needed from vehicle-based computation and prediction in the control loop. I like to think that I contributed to the cancellation of some foolish projects.

How did I do my research? The behavior of human operators was not sufficiently well known to predict their performance theoretically, though I read all of the literature and tried. I had to run experiments. I tested human operators both on computer simulations of vehicles and real vehicles of various configurations. By varying the problems, delay times, and speeds and using different experimental subjects I was able to gather enough data to develop theory to match it. I was then able to test this theory in other situations, and happily it worked.

Much theory is developed by extending existing theory, using experimentation to ensure that the extensions are valid. A professor friend of mine who supervises a large number of Ph.D. students has an interesting approach to research. He maps the work that has been done in his field as a function of the various quantities of interest. The results show certain islands of understanding and a great sea of ignorance. He encourages his students to do work that overlaps at least one of these islands but contains a reasonable amount of sea. The island gives them a starting point and the sea leaves them room to do their own work. The overlap will either agree or disagree with previous work, and either outcome will be interesting.

Experimentation is performed extensively during research and development in order to increase understanding and establish feasibility. Some of these tests can require very extensive equipment. The National Full-Scale Aerodynamics Complex at Ames Research Laboratory (see figure) consists of two wind-tunnel test sections that share a common drive system and an outdoor aerodynamics research facility. The two test sections have a cross section of 40 ft × 80 ft and 40 ft × 120 ft. This allows the testing of large (often full-scale) models. The tunnel cross sections are so large that the obtainable velocities are rather low (115 mph in the large section and 350 mph in the "smaller" one). However, the purpose of the tunnels is to test low-speed characteristics of high-performance aircraft (an important problem, since airplanes designed to go fast sometimes do not like going slow, though it is useful in taking off and landing). The tunnels also test high-performance helicopters and advanced rotorcraft (aircraft that have the hovering abilities of helicopters and the high-speed abilities of conventional airplanes). The complex is also used for fundamental research on large-scale fluid flow and

NASA Ames Research Center's Full-Scale Aerodynamics Complex.

acoustical problems, in which case it comes close to being a large scientific instrument.

The facilities, despite their relatively low-speed capabilities, are impressive achievements. They cover twelve acres, the closed circuit of the 40 × 80 ft tunnel is one-half mile in length, and the maximum air flow through the test section is 63 tons per second. The air intake for the 80 × 120 ft tunnel (which is a straight-through tunnel rather than a circuit) is 130 ft high and 360 ft wide (approximately the size of a football field). The drive system consists of six fans, each 40 ft in diameter, driven by motors with a total of 135,000 horsepower, which consume 105 megawatts of power. This experimental facility represents a huge investment, and it is safe to say that if theory alone were adequate to design airplanes, such facilities would never have been built.

Even if theory exists and is known, experimentation is necessary to bridge the gap between the idealized world of engineering theory and the complexity of reality. Indeed, theory and experiment are two parts of a continuum. Occasionally, theory alone suffices. Sometimes one must depend entirely on experimentation. But usually both are needed to solve the problem. The process of mathematical modeling requires that the world be simplified so that the equations can be handled. Approximations are made and certain quantities may be omitted in order to make the mathematics more reasonable. For example, it is much easier to work with relationships that are "linear." This means that quantities have a directly proportional relationship to each other: stress is proportional to strain, acceleration is proportional to force, current is proportional to voltage, and so on. Stress is proportional to strain in the elastic zone, which is the region in which we can deform something and see it return to its original state when we release the load. Fortunately, we tend to work most often in the elastic zone, but not always. Acceleration is proportional to applied force if no friction is present. Current is proportional to voltage if the wire is big enough, and so on. Yet it is often the case that things are not quite elastic, a bit of friction is around, and the wire is not as big as you thought. Therefore, designs when built often do not act quite as they are supposed to. In fact, not yet in my life has a prototype of something I have designed worked as I thought it would. Often, they have not worked at all. For instance, mechanisms "bind." We all know what that means. It usually happens because of a combination of elastic deflection in the device and friction. It is very difficult to predict binding analytically. I have designed many ingenious and sophisticated mechanisms that initially operated so poorly that they would have been funny had I not designed them. The theories I applied simply had not taken account of everything that was happening in the actual mechanism.

One of the difficulties in predicting the behavior of complicated technological products during design is the complexity itself. Engineers have become much more sophisticated at handling complexity over the past 100 years or so. Computers have made it possible to consider a very large number of quantities. Mathematical approaches such as operations research specialize in large numbers of equations with large numbers of unknowns. However, if the complexity is high enough, error will occur, because of the number of assumptions that must be made, accidental or erroneous omissions, or the above-mentioned difficulty in applying the theory. These errors are caught with experimentation.

I encountered this immediately after graduate school when I was assigned to a group involved with controlling the temperature of spacecraft subsystems. I had taken some courses in heat transfer, which is a discipline concerned with the mechanisms through which heat energy is transferred and temperature is distributed. They were reasonably straightforward. There are three basic mechanisms of heat transfer: conduction, convection, and radiation. If you straighten out a paper clip and then hold it by one end and stick the other end into a candle flame, you will experience discomfort by means of conduction. The heat energy of the flame is transferred directly to your finger through the agitated motion of the heated molecules in the paper clip. The amount of heat transferred is directly proportional to the cross-sectional area of the paper clip wire, the temperature gradient along it, and a constant called the thermal conductivity of the material, which is experimentally determined. Pretty simple!

When heat is transferred by a gas or liquid rather than a solid, the process is called convection. Your furnace makes use of convection, because it heats the air in the room. The theory is a bit more complicated to use than that for conduction. The amount of heat transferred is again proportional to cross-sectional area, to a temperature difference—this time between the hot surface (a radiator, say) and a reference temperature in the fluid (in this case, air)—and a quantity called the average unit thermal convective conductance. Unfortunately, this last quantity depends on the geometry of the surface, the physical properties of the convective fluid, the velocity of the fluid, the temperature difference, whether or not the fluid is changing phase, and many other factors.

Heat energy can also be transferred by electromagnetic radiation. This form of heat transfer is responsible for the pleasure of a fireplace. Fireplaces are generally poor at heating by convection and are therefore criticized by energy experts. Indeed, they are capable of convecting more heat up the chimney than they supply to the room. But fireplace advocates (like me) argue that fireplaces are more satisfying when the air is colder anyway, because the joy is in roasting the body through direct radiation, not in simply warming the air in the house through convection. One should not expect a fireplace to heat a room, anymore than one would expect to increase the temperature of the atmosphere appreciably with a campfire. I admit to being energy inefficient in my quest for radiated energy, but I atone for my sins by keeping the house bracingly cold (except for the fireplace).

The net rate of heat energy transferred by radiation between two "black" bodies is proportional to a number called the Stefan–Boltzmann constant, multiplied by the area of one of the bodies, times a quantity called the "form factor," which is the fraction of the total radiant energy leaving that body which impinges upon the other, times the difference between the fourth powers of the temperatures of the two bodies. Unfortunately, real bodies are not black. They emit and absorb at different levels depending on their surface characteristics and the radiation frequency. In addition, they reflect.

In my classes, we applied these relationships to problems that typically contained a few simplified bodies that were either black or had some simple "grayness." If there were corrections to be made to the relationships, they were given to us. I had even had a bit of experience in controlling temperatures, but always with motors and electronic equipment that transferred heat energy to the air, which then rose, leaving nice cold air to take its place.

Designers of unmanned spacecraft prefer not to rely on fluids for heat transfer, because there are none in space and any fluids taken into space for purposes of temperature control can be lost through leakage. Once unmanned spacecraft leave the atmosphere, heat transfer is accomplished by conduction and radiation. Some control of temperatures is done actively, by use of bimetallically controlled louvres, conduction switches, and other such mechanisms. But mechanisms are heavy and can also fail. In the interests of reliability, it is preferable to control temperatures by the surface finishes of the parts, the conduction paths, and shielding from the sun when necessary. This is quite possible. A thin gold plate in deep space at the distance of earth from the sun and oriented perpendicular to the sun's rays would come to equilibrium at approximately 405 degrees F, since gold likes to absorb energy at short wavelengths (from the sun) and does not like to emit it at long wavelengths (infrared). It therefore becomes hot. But if the plate were painted white, its temperature would drop to −18 degrees F, since white paint likes to emit at long wavelengths and is less happy about absorbing at short ones.

Unfortunately, the spacecraft I was to work on was very different from the type of classroom problem I had studied. It consisted of hundreds of objects having shiny surfaces instead of black and of such irregular shapes and orientations that form factors could not be accurately calculated; it also had a myriad of joints having conductivities difficult to predict. When I went to work at JPL, computers were not as ubiquitous as they now are, so even if one could define all of these quantities, the computations resulting were, to say the least, expensive and exhausting to perform.

Experimental approaches saved the day. For instance, in order to quickly approximate form factors from models of proposed spacecraft, some of the engineers invented a little instrument they called a form factometer. It was a small polished metal hemisphere with a pin projecting from the top. It was engraved with a pattern so that form factors could be determined by placing the device at the point of interest on the model, viewing the hemisphere along the pin and counting the squares filled by the image of any other part of the spacecraft. A great help. As another example of experiment, a full-sized temperature-control model of the entire spacecraft was built that was identical to the real one from a heat-transfer standpoint. Structural parts were given their proper surface finishes, electronic power was simulated by electrical resistance heaters, and the effects of solar panels were simulated. This was tested in a chamber that simulated the sunlight, cold radiation characteristics, and vacuum of space. A cross-section of the present JPL space simulator is shown on page 160. It consists of a stainless steel cylindrical vessel 27 feet in diameter and 85 feet high that can be evacuated to simulate the vacuum of space. The walls and floor are lined with cryogenic shrouds that can be controlled over a range of −320 degrees Fahrenheit (to simulate the radiation sink of space) to +200 degrees Fahrenheit (to simulate a planetary body). At the top is a solar simulation unit capable of providing a collimated beam that can be varied up to an intensity of twelve times that of the solar radiation at

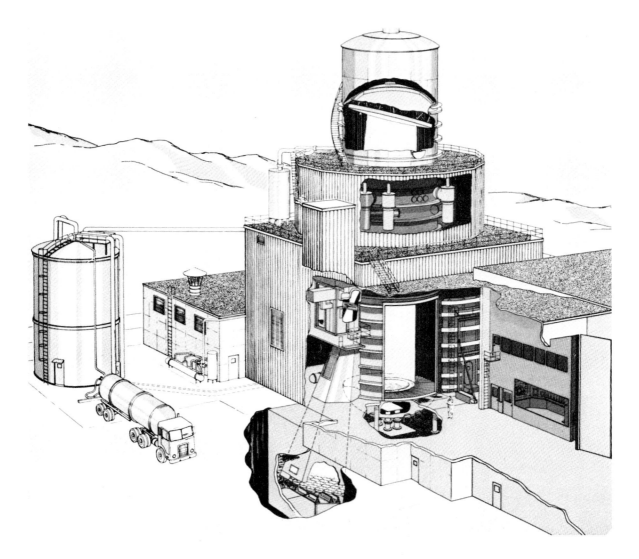

Caltech Jet Propulsion Laboratory space simulator.

the distance of the earth's orbit. The next figure shows a portion of the Galileo spacecraft inside the chamber. Galileo's mission is to explore Jupiter and its moons in detail. The spacecraft was launched in 1989. It is to pass Venus once and earth twice to gain velocity by taking advantage of the planetary gravities. It will reach Jupiter in 1995, releasing an entry vehicle as it approaches. It will then enter an orbital trajectory, first of all monitoring information from the entry vehicle and then collecting information from the planet and its moons.

After building many spacecraft, JPL's engineers have developed a better sense of the uncertainties involved and more conservative design approaches. Computers can handle a very large number of bodies. If they know the geometry of each body, they can calculate all of the form factors, making such things as the form factometer no longer necessary. Unfortunately, information is never complete as to the surface finishes of all the bodies, the conduction paths between bodies, and other such physical parameters. The differences between theory and reality can add up in strange ways if the number of bodies is large enough. Testing is necessary and is still done. It is often possible to rely on tests of portions of the system and less complete models of the entire system, but both theory and test are needed to adequately predict behavior.

Temperature control was not the only rude awakening I received from my schooling when I went happily off to aerospace. For instance, I found that I had studied static structural behavior much more than dynamic behavior and once again had dealt with simple models. The most severe environment a spacecraft experiences occurs during boost. The violent and widely varying vibration combined with acceleration is enough to foul up any straightforward isolation system one might design. The vibration lasts long enough for every part of the spacecraft to come to a fully developed resonance. This can be very bad news, because structural members in resonance can see many times the motion of the input. That can happen in earthquakes. Structures "tuned" to the frequency of the earthquake are in trouble.

Once again, complexity overwhelmed theory, and experiment was necessary to take account of reality. The figure on page 163 shows the Galileo spacecraft during vibration testing. Accelerometers are mounted to critical

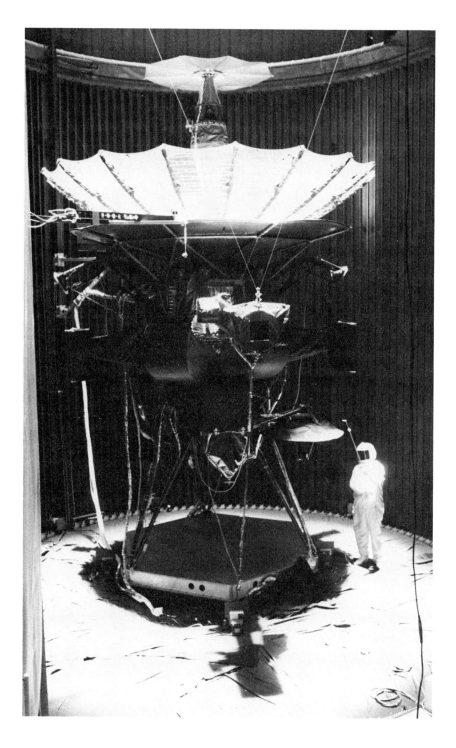

The Galileo spacecraft in the
space simulation chamber.

Vibration test of the Galileo spacecraft.

Development, Test, and Failure   163

parts of the spacecraft and it is mounted to a shake table, which is simply a large surface driven through an amplifier like a giant loudspeaker. The JPL vibration unit has a 192 kilowatt amplifier capable of frequencies between 5 and 3,000 cycles per second (Hertz) and, in case you wonder, is capable of playing music. Tapes of the desired vibration are "played" directly into the hardware, and data are taken from the accelerometers as well as by means of movies taken under strobe lights synchronized so as to slow down the motion and allow visual analysis. Once again an expensive procedure, but necessary because the complexity of the hardware exceeds the ability to predict its behavior on the basis of theory.

In the engineering of a product, a conflict often arises between those who prefer to experiment and those who specialize in theory. When some unacceptably large motions were discovered during testing of a spacecraft that was being built on a tight schedule, the engineers doing the testing wanted to put more damping into the area with the problems, since damping reduces the amplitude of resonance motions. That is why cars have shock absorbers. Without them your car would bounce repeatedly after hitting a bump and could resonate dangerously if you drove a particular speed over regular rough roads. The shock absorbers "damp" out unwanted motion. One of the more experimentally motivated engineers, who incidentally was a car freak, designed a couple of simple light-weight dampers (concentric tubes with grease between them) which did an excellent job of controlling the local motions. But this fix was strongly opposed by the structural analysts because their computer programs assumed that damping was evenly distributed throughout the structure. After a brief and vicious political battle, the dampers were tried and they worked, although the victory was difficult to celebrate because of the snarling of the structural engineers in the background. In this case, the high priests of theory actually tried to block a necessary development.

A good example of the type of experimentation involved in the development of a complex product is the "debugging," which is so integral to the computer business. As we realize by now, no matter how good the programmer, complicated software programs have errors. But hardware also has its problems and must be "debugged." The following anecdote is from Tracey Kidder's *Soul of a New Machine,* which is about the building of a computer at Data

General Corporation.[1] The "flakey" was discovered very late in the development program, and its solution captures the irrational side of engineering.

But Gallifrey, the lead machine now, still wasn't all there. It was running all the toughest diagnostics, but failing occasionally on some of the lower-level ones. The Hardy Boys would leap for their analyzers, run the test again, and the failure wouldn't happen.

"A flake."

But where was it?

On October 6 the vice president, Carl Carman, came down to the lab as usual, and they told him about the flakey.

Carman is a man of medium height, in his forties, fair-haired, with skin rather pink from the sun—all in all cherubic-looking. He smiles like Alsing, mysteriously.

The ALU was sitting outside Gallifrey's frame, on the extender. Gallifrey was running a low-level program. Carman said "Hmmmmm." He walked over to the computer and, to the engineers' horror, he grasped the ALU board by its edges and shook it. At that instant, Gallifrey failed.

They knew where the problem lay now. Guyer and Holberger and Rasala spent most of the next day replacing all the sockets that held the chips in the center of the ALU, and when they finished, the flakey was gone for good.

"Carman did it," said Holberger. "He got it to fail by beating it up."

Even after a design "works," continued testing must be done in order to ensure that it fulfills all of its requirements. I was at one time involved in flight testing aircraft at Edwards Air Force Base. These aircraft had already been through tests by the companies that built them and were now being performance-tested by the Air Force. One reason that airplanes are tested at Edwards is that the climate is superb for flying—we used to talk about the "flyable skies" of the Mojave Desert. However, another reason is that the size of the base (300,000 acres located in a sparsely settled area) is compatible with the risk of early flight tests. As we slightly irreverent lieutenants used to say, "It's a great place to test airplanes because no one can see them fall down." During the time I was there, one very sophisticated prototype fighter plane did not recover from a spin, another failed to go supersonic, and an ejection capsule for a research plane proved to be a death-trap. Landing gear got stuck, vibrations from weapons caused flame-outs, aerodynamic flutter showed up where it should not have existed, and control responses seemed to require incessant tweaking. Some of the testing bordered on the humorous. I recall a test performed on our high-speed track to ensure that the windshield of a particular airplane would not be adversely affected if the airplane hit a bird. Since it is not easy to fly around and successfully hit birds, a mockup of the entire front section of the aircraft, including the windshield, was accelerated down a track on a rocket-powered sled. Over the track was carefully suspended the proper bird (carefully specified and mercifully dead). One of my favorite engineering memories has to do with this huge, roaring, speeding, technological construction hurtling across the desert at this miserable dead goose. But we who have since ridden in the many generations of this airplane should be glad that this test was performed (it was the Boeing KC135, which was to become the prototype of the 707).

Testing is also done to ensure that products will withstand their environment and have the desired reliability and lifetime. Environmental testing can become quite complex. Military equipment has very stringent environmental requirements. When aircraft completed their flight testing at Edwards, they went to Eglin Air Force Base in Florida for environmental testing. This base had entire hangars that could simulate any type of weather you would desire (or not desire). Airplanes were subjected to the entire gamut, from noontime in the summer desert through hurricanes and sleetstorms to midnight in the winter Arctic.

Reliability and lifetime are becoming much more important as consumer expectations and liability costs rise. Life testing is difficult, especially in cases of high reliability and long life, because of the quantities and times involved. It is possible to accelerate life testing by increasing duty cycles, the severity of the environment, or modifying other factors that tend to make products fail. However, one must then make assumptions about the relationship of the test situation to the real one. Often it is impossible to avoid duplicating actual operating conditions. One of my students, who was a notable car buff, once found what he considered the ultimate summer job. It consisted of driving automobiles over test courses to measure reliability and lifetimes. He had not considered the fact that he would drive automobiles around these courses for eight hours a day and that the courses tended to wear heavily on either the mind (endlessly driving around a regular oval) or the body (Belgian blocks, ruts, gravel, bumps ingeniously devised to maximize wear on the suspension). When he returned to school, he was less of a car buff.

Once a design is completed and manufacturing begins, many tests are performed to ensure that the manufactured materials adhere to specifications. These can vary from simple testing performed to ensure that manufactured parts are within tolerance to the flight testing given to completed aircraft. In the case of small manufacturing runs, standard gauges and testing devices are typically used to inspect some or all of them to ensure that they are the desired geometry. For large numbers, the testing is often automated. Many integrated circuits are formed on a single "wafer" of silicon, and each device may have from 10 to 500 contact pads. A machine called a "wafer prober" automatically makes electrical contact with each of these pads and checks each device on the wafer for electrical performance. The machine can be programmed to carry out desired checking routines and modified for various contact-pad configurations.

This type of testing falls within the purview of industrial engineers, who talk of quality control and quality assurance. It involves the application of statistics. How many samples of the product must be tested for what lengths of time to be able to say that the product has a reliability of 99.9 percent? Obviously one would not like to test more than necessary, especially if the testing harms the part, but one wants to test enough. What does one do about testing if only a few, or just one, of the products will be manufactured (as with the space shuttle or space telescope)? These are all questions that are susceptible to mathematical treatment, but the answers do not always turn out right, as we know from hindsight. Quality assurance goes beyond statistics into issues such as the motivation of people to do high-quality work. The attainment of improved quality is a universal goal of U.S. industries at the present time, and the proper design and implementation of testing is an integral part of it.

One of the older philosophical debates, which still rages, has to do with the testing of products that will be used in extremely critical situations. In the case of spacecraft, for instance, how severely should the flight spacecraft be tested? One school of thought says that identical models should be tested severely, but the flight models should be given gentler tests in order to avoid damaging them. The other school of thought says that the flight models should be thoroughly tested. Most engineers belong to the latter school of thought; yet many people's instincts cause them to join the former.

A good argument for thorough testing has to do with the character of failure. Many components fail early in their usage, probably because of imperfections introduced during manufacture. Then the likelihood of failure decreases and stays low for a long time. The failure rate increases again as the components approach their service lifetimes. This behavior is an argument for thorough testing. In fact, high-reliability electronics are often "burned-in" before delivery, which means that they are run either in normal operation or at an accelerated level until they are through the early part of their lives.

For sophisticated products, testing may continue during the service of the product. The periodic checks on automobile emission equipment in some states are an example of this. You should be glad to know that aircraft equipment is periodically tested. Your lap-top computer is able to test its own battery charge and tell you if it is too low. The reason you have to wait for your computer to speak to you after you turn it on may be because it is testing itself.

Much testing is for the purpose of catching plain old mistakes, or at least things that look like mistakes in retrospect (a common human perception). The press and media love to tell us about such things after the causes of accidents are found. Most engineers would prefer to find their mistakes before the press does, through testing. Even after all of this testing, failures do occur. Your engine stops running, your washer will not drain, airplanes crash, structures collapse, and nuclear plants release radioactive material. Why is this? This question is becoming more and more important as projects become larger and potentially more damaging and as public expectations rise. The legal and social aspects of failure will be discussed in Chapter 10. But let us think a bit here about the technical side.

One of the ironies of engineering is that progress involves failure. If engineers restricted themselves to exactly duplicating past projects that had not failed, one might think that failures could be avoided. This is not the case, however, since variations in environment, materials, the production process, and usage would add uncertainties. In addition, we would still be using stones to kill animals for food.

When a failure occurs, the source of failure is often identified, and in hindsight it usually seems as though the cause should have been caught in the process of design and development. Sometimes, these failures occur because the designers and all of the people who checked their work simply did not think of a complication or catch a mistake. Sometimes business decisions do not allow sufficient time and resources to work to the necessary level of detail. But in other cases the effort or expense required to discover the cause or to design so that failure could not occur is simply unreasonable, given the likelihood of failure. If the failure were caused by a crack in a structural member, what would have been involved in examining all of the members for cracks nondestructively? If a human manually overrode an automatic shutdown, what would have been required to make such an act impossible? Sometimes, the effort required would have made the product unaffordable.

One's instincts often tell one that a large amount of redundancy should be provided, so that if critical components fail, others can be substituted. But it does not take one long to realize that the necessary detection and switching mechanism may add more unreliability to the system than reliability. Aircraft must be designed with relatively low safety factors or they simply could not fly. An airplane that could not crash is simply impossible to conceive. In addition, some physical phenomena are extremely subtle. A good example of this is fatigue, which is a particular problem with parts that must endure large numbers of loading cycles, such as rotating shafts or flexing structural parts. In fatigue, a material defect, a discontinuity in the shape of a part, or an overly high stress will start a miniature crack. The alternating load cycles will then cause this crack to enlarge slowly, until the intact material is inadequate to carry the load, at which time failure will occur. The cracked portion often has a smooth appearance due to rubbing together of the surfaces. The final failure surface often shows the grain structure and is therefore rough in texture. My grandfather and his friends referred to this type of failure as crystallization, in the erroneous belief (which still exists among many people) that somehow the rough texture meant that the metal had converted to a weaker crystalline structure.

Special problems arise in very sophisticated systems. An example was the explosion of the Space Shuttle *Challenger,* which kept us glued to the television screen in macabre fascination, watching again and again the destruction of the machine and its crew. The Shuttle is extraordinarily sophisticated and extremely complex. The design makes use of many features that had never previously been tried, such as the insulation tiles that protect the main vehicle during reentry into the earth's atmosphere and the reclaimable solid booster rockets. A large amount of experimentation and testing was required in its development and in proving the final version.

By the time of the ill-fated *Challenger* flight, the public and many people involved in the government and the Shuttle program itself had begun taking the safety of the Shuttle for granted. This was a mistake, for riding large tanks of explosives into space will not be routine for some time to come. However, it is a normal response of humans, and confidence had reached the point of using the shuttle to give rides to nonessential passengers and of performing missions that perhaps could have been more safely accomplished with unmanned booster rockets.

Partly for this reason, the explosion horrified the country, and the President, who was himself deeply troubled by the incident, appointed a commission of impressively qualified people who were not involved in the Shuttle program to find out what had happened. The detailed reason for the accident was isolated surprisingly quickly. A seal at the aft end of one of the solid booster rockets had failed. There had been rising concern among engineers about this seal before the flight. It consisted of a metal interconnection with two O-rings in the leakage path. O-rings look like slender rubber doughnuts and reside in annular slots in cylindrical joints which must contain a pressurized fluid. The fluid, in attempting to escape, deforms the O-ring to better seal the joint. Since O-rings are resilient, they can accommodate small irregularities in the joint and small motions. You may have encountered them in your kitchen sink spout, in bathroom faucets, or in your local hardware store.

Unfortunately, the critical resilience of O-rings depends on their temperature, and the launch was on an unusually cold day. Ice was found on the launch pad that morning, and the air temperature was 36 degrees at the time of launch, 15 degrees cooler than any other launch. This temperature dependence had been a concern of engineers during the development of the booster and had been communicated to company managers and NASA officials. In fact, an engineer had attempted to have the *Challenger* launch delayed because of this low temperature but had been unsuccessful.

Analysis of photographs verified the reason for the explosion. During the first few seconds of launch, dark puffs of smoke were seen emerging from one portion of the seal area at a frequency corresponding to the vibration of the structure at that point. The color of the smoke was indicative of burning grease and O-ring, and the puffs indicated that the leakage was occurring as the mechanical joint opened and closed slightly during the dynamics of launch. Just under a minute after launch, a plume of flame appeared from this same point and was deflected onto the main propellant tank (hydrogen and oxygen) and the rear strut that held the solid booster to the main body of the Shuttle. At the same time, the chamber pressure in the booster dropped, corresponding to a major leak.

A few seconds later, the color of the plume changed abruptly, indicating that the main tank had ruptured, adding hydrogen to the flame. Approximately 72 seconds after launch, the rear strut linking the solid booster to the Shuttle parted, permitting the booster to pivot around the forward strut. A second later, the entire aft dome of the hydrogen tank dropped away, releasing the load of liquid hydrogen and creating a thrust of some 2.8 million pounds that shoved the remainder of the tank toward the oxygen tank. At about the same time, the pivoting solid booster hit the bottom of the oxygen tank. The explosion followed immediately. The villain was obviously a joint design that was inadequate for the environment.

But was the design inadequate or was the launch environment too risky? Why had no heed been taken of early concerns, or of concerns expressed on the day of the launch? The commission spent much of its time on such questions. A large number of individuals and organizations were involved in designing and building the Shuttle and remained active in its maintenance and operation. Decisions were made hierarchically, and at each level judgments were made as to whether specific potential problems should be considered or not. There was increasing pressure to maintain Shuttle launch schedules in order to place high priority payloads into orbit.

Only one of the commission's findings demanded an improved and certifiably adequate seal design. Others asked for more centralized management of the Shuttle program to eliminate any conflicting interests of NASA Centers, better communication to gather all pertinent information on flight safety problems, a NASA safety organization reporting directly to the administrator, establishment of a firm and safe policy on launch schedules, and improved maintenance of critical components. In addition, the commission called for a safety audit of other critical shuttle subsystems, a provision for crew escape in the case of launch abort, and improvements in the landing characteristics.[2] Many engineers and managers associated with the *Challenger* disaster were transferred or retired, but none of them was held personally accountable. The seal was indicted, but it was a symptom of an organizational problem. The challenge of integrating large numbers of people and companies in an unprecedented engineering effort, meeting an ambitious schedule, and still keeping track of a myriad of details such as cold O-rings is immense.

Some causes of failure appear to be just stupidity when viewed in hindsight. However, in spite of our sophistication, our experience, and our access to powerful tools, we do not know how to predict or completely control the behavior of humans. They are not omniscient. They do make mistakes. Good engineers attempt to design products that are "people proof." However, when viewing failures in retrospect, one is often impressed at the ingenuity with which people manage to destroy the products of technology. One of my first design assignments had to do with increasing the safety of a high-speed milling machine built to fabricate a particular aluminum part. The cutting speed was so high and the aluminum so soft that the sound of the cutting was barely discernible over the high background noise in the area. To make matters worse, the teeth on the cutter could not be seen because of the speed of rotation. Operating the machine was repetitive and boring and required the operator to load the part into the holding fixture and then press a button to begin the movement of the cutter through the part. All of these characteristics resulted in an unacceptably dangerous situation.

I designed what I thought was the obvious solution. I installed another large button which also had to be depressed before the cycle would begin. I naturally assumed that if both hands were on buttons when the cutter began its travel, they would be safely out of harm's way. To my amazement, shortly after the machines were modified an injury happened. The operators were of course placing various objects on my new button to hold it down so that they could begin the cycle with one hand while their other was still loading the part and therefore in position to be sliced up by the tool. I then placed the button on a vertical surface, and it took a whole hour before the operators learned to tape it down.

A tragic DC-10 crash upon take-off from O'Hare Field in Chicago in 1979 got a tremendous amount of publicity that almost grounded the DC-10 permanently. The crash was due to the failure of a pylon that held the engine to the wing. Investigations of other DC-10s revealed cracks in their pylon attachment points. Fortunately for the DC-10, however, whose record since that time indicates that it is a superb aircraft, investigations showed that the cracks had been caused by a maintenance procedure which the designers had not, and probably could not have, anticipated. Maintenance people changing the engines were using a rather unsubtle tool called a fork lift. It was some time before the guilty maintenance procedure came to light.

The procedure was invented because it simplified the job of the maintenance people. But to the designers, it was a clear example of the difficulty of people-proofing a design. As another example, great care is taken in the design of electrical connectors for aerospace equipment so that only the proper male and female connectors can be assembled. In fact, there is a cylindrical shell around large male connectors that protects the pins as well as ensuring that male cannot be mated with male. However, I remember a case of a technician trying to do so hard enough that he was able to complete some wrong contacts and blow away some very expensive hardware. The only way to gain information about the way humans are likely to behave is through experiment, and even then one is likely to be unpleasantly surprised. I remember still another situation in which flight spacecraft hardware was being assembled in a controlled clean environment, where technicians wear white suits, masks, and caps and the air is filtered to remove all particles larger than a few microns in diameter. To our amazement, we found tomato seeds from someone's lunch in one of the pieces of equipment that was being assembled.

Much of liability law is an attempt to establish whether "reasonable practice" was followed in the engineering process. Let us discuss another particular failure to gain a better feeling for this. Your reactions will be, "It shouldn't have happened." But keep in mind that engineering is an occupation often calling for many people to do unprecedented things.

In the Hyatt Hotel in Kansas City in 1981 two elevated walkways, originally suspended one over the other, collapsed onto the lobby floor below, causing 114 deaths and some 200 injuries.[3] The original design called for these two walkways to be hung by rods suspended from the ceiling. Each walkway was to be built on beams, and the beams were to be held by threaded nuts on the rods, which were long enough to support both walkways. The post-collapse investigation concluded, first, that the rods were not as strong as they should have been according to the code. Since this was not a traditional design, that point was debatable, however, and given that building codes are quite conservative, the original design should have held the load. The load itself was not controllable, since any number of people could crowd onto the walkways; at the time of the collapse people were dancing on them, creating a dynamic situation that could have been structurally more damaging than if they had merely been standing still. But the original design, although perhaps marginal, should have been adequate for even these stresses.

The original design was modified, however, for ease in construction. In the original drawings, the details of threading the rods for the nuts and washers to hold the beams was not clear. The rods were very long, and it would have been difficult to thread the entire length and turn the nuts to the desired position. It would also have been difficult to fabricate rods that were larger in diameter at the levels of the walks so that just the larger portion could be threaded. Therefore, the design was modified as shown in the figure.

Two sets of rods were used, each threaded at both ends. One set suspended the upper walkway from the ceiling, and the other suspended the lower walkway from the upper. This allowed the rods to be transported and installed more easily. Is this approach structurally different? If so, how? The answer is that it is, but even engineers have to think a bit to see why. The load on the rods is the same in both cases, in that the upper set of rods is supporting the weight of both walkways and the lower set is supporting the weight of one. In the modified design, however, there is twice the load on the upper set of nuts and washers. In the original design, each set of nuts and washers is supporting one walkway. In the modified one, one set of nuts is supporting the lower walkway, but another set is supporting the weight of both.

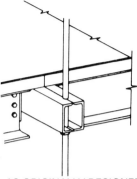

AS ORIGINALLY DESIGNED

In fact, the failure occurred when one of these bolts and washers pulled through the beam. Since this increased the load on the adjacent beam, the same thing happened there, and the failure propagated until the walkways collapsed. Should someone have noticed this detail before the walkways were built? Sure, but realize the complexity of a construction project of this extent and the number of details and new approaches that had to be handled in a relatively short time. In retrospect it was a mistake. Was it understandable? Yes. Was it forgivable? The courts decided that it was not, but through a long and involved process.[4] As of December 7, 1983, 397 civil claims had been filed and over $80 million had changed hands in settlements (another $20 million was being appealed). However, after a two-year study, no basis for criminal charges was found. A hearing judge for an administrative court found the chief engineer and his boss to be guilty of gross negligence, and they resigned from their jobs. Still later the state revoked their licenses. Yet their negligence was not in making the original error but in failing to catch it. The investigation, which was extremely thorough, apparently never did succeed in pinning the blame for the design change on an individual. The federal court was still reaching decisions as late as 1987.

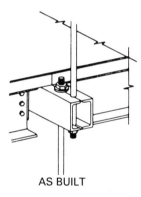

AS BUILT

Modifications in the elevated walkways at the Hyatt Hotel in Kansas City.

Think of other failures that you have heard of through the media. In the case of the Hubble Telescope and its poorly focusing mirrors, the decision was made not to do a complete simulation of the final assembly in operation because of the expense. However, a lot of very smart people took part in this project for a very long time. In hindsight, it looks as though the simulation should have been done. But was it a wrong decision or was the project simply so complex that the odds of something going wrong were quite high?

How about the so-called aging airframe problem that was in the news a couple of years ago? Many commercial airliners are quite old now, and the higher speeds of contemporary aircraft along with higher reliability and therefore a higher percentage of time in service have resulted in problems due to stresses and corrosion. The airframe of a 707 or a DC-8 works harder than the airframe of a DC-3. The problem has been identified, and airframe maintenance to repair and counteract damage is now under way. But should the problem have been prevented in the first place or detected before it happened? That is asking a lot of humans, no matter how competent. Experience with older propeller-driven aircraft had focused attention on the engines, the control systems, and almost everything else except for the basic metal structure.

The Three Mile Island and Chernobyl nuclear plants, the Bhopal and Stark incidents, the Exxon spill, and the exploding Pinto gas tank are a few more well-known "incidents" of failure. In each case, it was possible to look back and find error and pin down blame. However, all of these situations involved complex technological systems and humans, either in the design and construction of the systems or in the operation of them. Granted that they were all tragedies, and tragedies should not occur. But most of the participants were professionals, generally competent at their jobs, and I think concerned. Perhaps one can conclude that as long as we live with complicated technological systems and demand increasing sophistication and affordable costs, things will occasionally go wrong, despite wishes and efforts to the contrary.

8

Recently, my wife and I were given a tour of an automobile assembly plant. I have spent quite a bit of time in large assembly plants, but each time I go through one I am newly impressed. My wife had never been inside a plant of this type, and during the tour made a comment that I think represents the feeling of most people who encounter such an operation for the first time. She pointed out that the automobiles themselves are fairly trivial objects compared with the assembly plant. The technology required to manufacture and assemble them seems to dwarf the technology of the products themselves. The definition, design, construction, and debugging of an assembly plant is an extremely large-scale engineering job. The plant for the General Motors Saturn automobile project cost approximately $1.9 billion, out of a total project cost of $3.5 billion.

Although manufacturing and assembly (sometimes referred to as production) are often considered to be a separate step from design and engineering, in reality they are inseparable from the rest of the process. A design that cannot be manufactured and assembled well is obviously not a good design. Much of the challenge in design, especially in the design of complex objects or objects that are to be produced in large numbers, is in ensuring that the components can be economically manufactured and assembled into the desired final result.

Manufacturing and assembly functions are hidden from most of us and are generally taken for granted. How do you imagine the very common items in the list on page 178 are made? These cheap items obviously must be produced in large quantities by some inexpensive process. I sometimes use the list to torture engineering students, who have to admit that in most of the cases they do not know.

In recent times manufacturing and assembly in the United States has been taken for granted not only by the general public but by many people directly involved with the technological process. This complacency has cost us dearly in our ability to compete economically in the world. Many of our present problems, such as large trade and budget deficits, have to do with the fact that other countries have gotten better than we are at manufacturing and assembly. Why are manufacturing and assembly so important, and why did

Common manufactured objects

steel wool
aluminum beer cans
light bulbs
tennis balls
BBs
turbine blades for jet engine
rope
wooden pencils
nails
rubber bands
needles
mirrors
paper clips
toothpaste tubes
golf balls
matches
nylon thread
hollow metal doorknobs
resistors
nuts and bolts
clothes labels
wood screws
automobile crankshafts
wine bottles
typing paper

the United States lose the lead it once had over the rest of the world? Once again, history tells us some of the answers.

Until the middle of the eighteenth century, technology was craft-based, relying on the experience of individuals in designing and making products. The Industrial Revolution, with its steam engines, iron and steel, factories and cities, was a revolution in production. Tremendous increases in the availability of energy and the sophistication of machinery benefited first the manufacture of mechanically made products and the raw materials and construction industries, and later the electrical and chemical industries. Industries such as electronics and aerospace were in the future, but the means of economically producing precisely controlled parts and products from a wide variety of raw materials was well under way.

Think of the changes between, say, the sixteenth and the nineteenth centuries. Opposite is a drawing of a "factory" for making armor in the sixteenth century. Armor was probably produced in a more sophisticated manner than most hard goods. However, there were few machines and no such thing as standardization. If your armor broke, you did not pick up a replacement part at the local armor shop; you returned it to the maker. Standardization did not become a principle of engineering until the early nineteenth century. In the sixteenth century, the chemical industry consisted of individuals who manufactured soap, dyes, alloys of metals, and other such materials in small batches by hand. Although people knew how to make products through chemical reactions, they certainly did not understand the processes, and were a long way from the production of nylon cloth or urethane varnish. Construction, as it had been through most of time, was accomplished by means of hand labor, aided by animals and simple machines to lift weights and transport loads. The electrical industry did not exist.

By 1850 a tremendous change had occurred. Machine tools had proliferated and steam power had augmented wind and water power. Designers were realizing the advantage of interchangeable parts. In the United States interchangeable manufacture began with a government contract given to Eli Whitney in 1798 to manufacture 10,000 muskets, the parts of which were to be interchangeable. However, Samuel Colt and Elisha Root were still able to

A sixteenth-century armor factory.

dazzle attendees at the Crystal Palace Exhibition in London half a century later by placing baskets of revolver parts on a table and letting passersby assemble revolvers by taking a part from each basket.

The Crystal Palace itself was a good example of mid-nineteenth century production. It was designed by Joseph Paxton, a gardener who had previously confined himself to the design of greenhouses. His extraordinarily successful and magnificent structure was built with standard components: 6,024 identical iron columns, 1,245 identical wrought iron girders, 3,000 identical gallery beams, 293,655 identical panes of glass, 3¼ miles of pipe, 45 miles of standard sash bar, 600,000 cubic feet of timber, and assorted nuts and bolts. It was built and erected for a fraction of a cost of the competing designs; and after the exhibition was dismantled, the components were sold to other builders.

A revolution in chemistry paralleled the Industrial Revolution. During the eighteenth century the idea of a large number of "interchangeable" basic elements (replacing earth, air, fire, and water) was accepted and refined, the concept of the conservation of matter led to quantitative methods, the nature of gasses and the process of combustion were better understood, electrochemistry was studied, and improvements in atomic theory led to a much more sophisticated analysis of chemical reactions. By the middle of the eighteenth century, chemical industries were emerging in Europe. The ability to combine purification by crystallization, distillation, and the handling of gasses with furnace techniques led to plants dedicated to the relatively large-scale production of soda and potash, glass, soap, acids, dies, and fluxes. Increased industrial activity resulted in ever more products, which in turn stimulated other products. Stephen Miall, in his *History of the British Chemical Industry, 1634–1928,* describes this process as follows:

> The works of the chemical manufacturer tended to become larger and more complicated; he began to make soda, using common salt and sulphuric acid and other raw materials. After a time he started to make his own sulphuric acid by burning sulphur or pyrites; if he used pyrites, it was probably a mixed sulphide of copper and iron, and it was comparatively easy to make copper sulphate and ferrous sulphate from the

roasted pyrites. The process of making sodium sulphate produced large quantities of hydrochloric acid, and as nitric acid was required in the manufacture of sulfuric acid, the alkali manufacturer easily developed into a manufacturer of hydrochloric, nitric, and sulphuric acids, and various salts of sodium, copper, and iron. It was a very common development for the alkali manufacturer to use the chlorine he recovered so as to make bleaching-powder, and in this way he became a maker of calcium chloride and bleaching-powder, and as demands for them grew, he made other salts of sodium and calcium required in large quantities. The manufacture of all these "heavy" chemicals became in this way an involved process, in which one part was dependent on the others and almost every effort to prevent waste involved the manufacture in some new product.[1]

No computer control yet, but lots of action.

The photograph on page 182 shows a view inside a modern factory, one of Apple's Macintosh production plants. Notice that the change from the armor factory includes more than the product and the costumes. Machines are more prevalent than people. Apple has made a tremendous effort to design producibility into its computers and peripherals and to build production plants that are flexible, in that they can manufacture a wide variety of Apple products with the same equipment. In modern production plants computer information handling, control, and testing allow the accuracy necessary to produce products of very high sophistication at comparatively low cost. Modern chemical plants and construction sites show the same trend. Machines have picked up the physical work, computers have taken over control and information handling, and consequently the number of humans involved has decreased. But perhaps the most dramatic figures are from agriculture. In 1870 in the United States, well over half of the labor force was working in agriculture. By 1970 this had dropped to 3.5 percent of the labor force, though the amount of food produced per capita has increased markedly, as our exports will attest.

Between 1890 and 1930, total U.S. productivity more than doubled. Between 1930 and 1970 it increased by another factor of three. In a sense, our lives have been affected even more by incredible advances in our production

A modern Apple Computer factory.

abilities than by improvements in the products themselves. Tape recorders are marvelous, yet they can be purchased for the price of a single meal in an expensive restaurant. Most of us would have far less difficulty preparing an equivalent meal than an equivalent tape recorder.

From the turn of the century through World War II, the United States led the world in ability to manufacture products of high quality at low cost. Most people would agree that the decisive factor in World War II was the United States' production ability. The war simply furthered our lead. Our industrialized competitors were set back by the war, either through damage, social dislocation, or economic strain. As an indication, in Europe approximately 30,000,000 people were killed, and the industrial and agricultural production of the European powers were decreased by more than half. The United States lost 400,000 lives, the index of industrial production nearly doubled, and the Gross National Product increased from about $90 billion to about $170 billion. In addition, we increased our technological lead in production, our pool of skilled people, and our industrial plant even further. Our recovery from the war was almost instant, and the subsequent rapid expansion was based partly on a large world market for our products.

Other countries, envious of our economic prosperity, set about upgrading their own production ability, drawing on our experience when possible. Japan and Germany caught up with us rapidly, in part because their old plants had been severely damaged during the war and new ones had to be designed. U.S. industry was making a lot of money, and the United States experienced 30 years of unheralded prosperity and global dominance. But our approach to production was stagnant.

About 1970 we began to realize that we were in trouble, as basic industries such as steel and automobiles began to lose their market to competition from other countries. We had taken our superiority for granted and grown fat and happy on our past record. U.S. industry is now justifiably concerned about production and is frantically trying to upgrade its capability. Advanced production is essential to the health of industrialized nations. Without it, the best inventions in the world have little value.

Let us consider a few aspects of production. First of all, we will look at some processes of manufacturing and assembly and the machinery that is required. Then, we will consider changes in the "factory," or the environment in which production occurs.

The basic processes of manufacturing range from very old to very new. A very old one is cutting. Most adults have cut vegetables with a knife, and probably wood and metal with saws and drills. We see soil being cut with bulldozers. People initially cut things by hand with sharpened rocks, and then later with bronze, iron, and steel knives, axes, plows, and drill bits. Now individuals also cut with heated wires, flames, and lasers.

By the middle of the nineteenth century modern machine tools had been invented, and cutting was no longer necessarily done by hand. The lathe, drill press, milling machine, shaper, and other familiar cutting devices had been invented. In Chapter 1 we saw a lathe designed in the late eighteenth century. These older machines may seem primitive compared with the computer-controlled lathe shown in the figure here, but the principles are the same. Old and new lathes cut metal by rotating the workpiece and moving a tool along it. Like present machine tools, the early ones were made of steel, built to high precision so that they were capable of producing parts to high precision, and powered by an energy source other than the human. The tools were mechanically guided, which allowed the machining of geometrically true surfaces to high precision. If you have ever run a wood lathe, you realize that it is challenging indeed to hold the tools by hand in such a way as to attain precise dimensions. A lathe designed to shape metal makes this task easier, granted that machines to shape metal are more complicated and more knowledge is needed to hold the part properly, choose the proper tool, cutting speed and feed, and measure the result.

However, the modern metal lathe in the figure is also very different from its early ancestors. The materials used to make it and its tools are far more sophisticated. It is much better designed, in part because designers now have more experience but also because they have access to such things as computer-based structural-analysis programs. The machine's components are far more accurately made, allowing greater precision in the finished products. Finally, it has many more features, can be operated at much higher speeds, and is controlled by a computer. The computer selects the appropriate speed (the relative velocity of the tool and the part as one moves past the other) and the

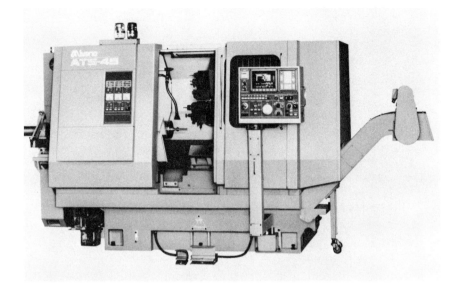

Miyano Machinery's computer-controlled lathe.

appropriate feed (the amount of metal removed as the tool and the part move past each other), selects the proper tools, and then controls the motion of the tools so that the part reaches the desired final shape. Twelve tools are held in a rotating turret that can cycle from one tool to another in less than a second. Another twelve tools can be held in an automatic tool-changer magazine. Seventeen of these tools can be rotating. A secondary spindle can grip the part to allow turning operations on the end that is held for initial shaping operations. Spindle speeds to 4,000 rpm are possible, and available accessories include systems to feed raw stock into the machine, parts catchers and conveyors, and chip conveyors.

Machine tools are a lot simpler than they seem to the uninitiated. One of the more popular courses at Stanford is entitled "Manufacturing and Design." In this class, the students design and build projects out of metal. Most of them have had no experience with machine tools before the course and are given only simple introductions. They find that they can figure out how to

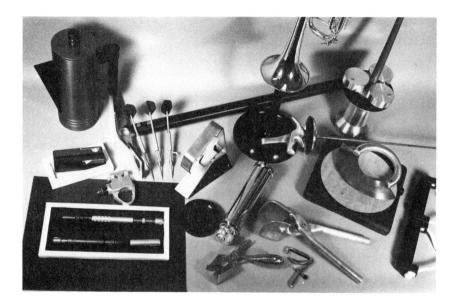

Metal products designed and built by Stanford University students.

run these complex devices, and that once they can do so, they can make most impressive products. The above figure shows a few student projects. This course not only imparts knowledge and skill but adds greatly to the confidence of the student and demystifies the manufacturing processes.

But cutting is not the only way to shape material. Another way is by molding, or casting, in which a fluid material is allowed to harden inside a mold. The material may be molten metal, it may be a compound that will harden through chemical reactions (a polymer or concrete), or it may be something more exotic, such as fine metal powder that is bonded together by heat. Ever made candles at the beach by digging a hole and pouring in molten wax? That is a simple form of sand casting, one of the most widely used molding processes. Many metal parts having a rough texture that looks somewhat like fine sand were cast in a sand mold. Parts of the lathe shown earlier such as the bed and the tailstock structure were undoubtedly sand castings.

There are many other forms of casting. In die casting, the material (typically zinc or aluminum) is forced into a steel mold under very high pressures (20,000 psi). A large and precise machine is necessary to control the forces involved in handling the heavy steel molds and holding them together as the metal is injected. (The force needed to hold together a mold with a cavity 6 inches high and 6 inches wide would be 720,000 pounds.) However, with die casting the finished part is extremely accurate and has a very fine finish. The body of the carburetor in your car is a die casting. The making of the molds is often a complex manufacturing process in itself, and the cost may easily run into tens of thousands of dollars. Other types of casting (shell mold, centrifugal, investment) simply differ in the nature of the mold, the process by which the material is introduced into the mold, and the environment in which it is allowed to cool. (Centrifugal casting, for instance, spins the mold and the molten material in order to ensure that porosity is minimized.) One of the more interesting forms of casting is continuous casting, or extrusion. It is the process used to manufacture pasta and all of the long, complicated aluminum shapes one finds in the hardware store (threshold, decorative angles, and so on).

Many processes used to shape plastics into the desired form, such as injection molding (the equivalent of die casting—used to make such things as plastic model kits and telephone bodies)—qualify as molding. Others, such as fiberglass layup (the process by which boats and Corvette bodies are made), are hybrids. However, to the extent that they allow a liquid material to harden to a final form, they share commonalities with the other molding processes.

A third category of manufacturing parts is forming—in which materials are plastically deformed by the application of the proper force. They may be bent, pounded into shape (forged), pushed into a die of the proper shape, or whatever. The process may be done either hot or cold. The armor in the figure on page 179 is being shaped by forming—by hammering it from sheet metal. The cabinets for your washer and dryer and the body parts of your automobile are formed by presses of very large dimensions. When I was an undergraduate, I used to love to take field trips to the Norris Thermador plant in Los Angeles, because the steam-operated forging hammers were so large that when they

Computer-controlled welding machines on a Ford assembly line.

hit the part that was being formed, the cars in the parking lots would bounce on their springs. Steel bathtubs, cartridge cases, and shovel blades are made by forming. The manufacture of beer cans and flashlight battery cases involves an interesting method of forming called impact extrusion. A small cookie of material is put into a mould and a die is smashed into it at high speed. The material climbs up around the die, and voila! a beer can.

Joining—the fastening together of materials—is one of the more critical operations in manufacturing. Mechanical fasteners may be used, such as nuts and bolts, nails, screws, or rivets. Materials are also joined together by welding, by using intermediate materials such as solder, or by using adhesives.

The figure opposite shows two computer-controlled welding machines (robots) capable of automatically joining the various complex sheet-metal panels that comprise an automobile body. These are two of 51 robots and other automatic machines that make 97 percent of the 4,000 spot welds on the Ford Aerostar minivan.

Although many fastening techniques have been around a long time, many changes are under way. For instance, machine-driven screws are replacing nails in construction because they give a stronger joint that does not loosen over time. Your hardware store is filled with super-glues (cyanoacrylate), fast-drying epoxies, and other formulations that were not there even five years ago. In the last 20 years an explosion of new materials has had a major impact on manufacturing and assembly. The ones at the bottom of the adjacent list are primarily used in the construction industry. Those at the top are the traditional ones. In the middle are some, especially the polymers (plastics), that are having a drastic effect on the nature of produced goods.

Many other processes of shaping material exist that do not fit into these categories. Grinding removes material by means of hard particles held in a matrix. More modern examples are electrical discharge machining, in which material is eroded through use of a carefully controlled electrical discharge; magnetic forming, which uses a magnetic field to force material into a die; metal spraying, used to build up materials by spraying on a molten metal; and laser cutting. Etching, plating, and use of various chemical coatings are also very important, though they often cause great headaches in production. Such processes are widely used in the electronics industry.

Integrated circuits or "chips" require that various materials be deposited upon the base material (in this case the semiconductor silicon) and etched into geometries that will provide terminals and conduction paths. Their manufacture also involves various chemical treatments to change the composition of portions of the circuit. We will take a moment to look at the manufacture of these devices in a bit more detail, because they provide a good illustration of modern production processes.

The first problem is to secure a large and very pure single crystal of silicon. The raw silicon must be purified by being melted in the presence of carbon, pulverized and allowed to react with anhydrous hydrogen chloride, and finally

Common materials used in manufacturing

Metals
  iron
  steel
    (iron with carbon)
  aluminum
  magnesium
  copper
  brass
    (copper with zinc)
  bronze
    (copper with tin)
  titanium
  tungsten
  lead
  tin
  gold
  silver
Polymers
  thermoplastics
  thermosetting plastics
  composites
    (fiberglass, honeycomb,
    filament reinforced)
Adhesives
  (often polymers)
Coatings
  (often polymers or metals)
Semiconductors
Glass
Concrete
Wood
Asphalts and other petroleum
  products

deposited onto rods of already purified silicon. The pure silicon is then melted and single crystals are formed by a specially built machine that slowly "pulls" them from the melt in a vacuum environment. These crystals are typically several inches in diameter and several feet in length.

The crystals are then ground to their final diameter and a locating flat is ground along their length at a precise angle relative to the direction of the molecular structure of the crystals (determined by an X-ray technique). The large crystal is then sawed into wafers with precisely controlled surface orientation and thickness. These wafers are then etched to remove any surface damage and contamination and polished to provide an adequate surface for photoengraving.

The process of manufacturing functioning integrated circuits from these wafers entails depositing other materials, growing thin crystalline layers (epitaxy), implanting ions, oxidizing, diffusing, etching through various layers, and coating. Multiple integrated circuits are produced on each wafer at once, and a process called photolithography is instrumental in doing so. This process begins with a large layout of a pattern that is desired on the chip. The layout is usually produced by computer and is then photographically reduced to the desired size. The image is projected onto a surface that has been covered with a photoresist which changes its chemical nature where it is struck by the impinging radiation. A developing solution then dissolves either the exposed or unexposed resist (both types are available), leaving the material locally susceptible to an acid etch. An etching process can then be used to create a mask, which can in turn be used to allow direct photolithography on the chip or to control deposition and coating patterns. In the past, light has been used for the photolithography radiation, although present small sizes are requiring more use of short wavelength radiation such as X-ray.

The figure at the top of the next page shows the steps needed to form a simple metal-oxide-semiconductor (MOS) capacitor. The bottom figure gives an indication of the complexity of the fabrication sequence for a portion of an n-channel metal-oxide-semiconductor (NMOS) logic circuit. I will not go through the details, but I include this to show the closely controlled processes needed to build integrated circuits. If one remembers that dimensions on the order of one millionth of a meter (a fraction of one ten-thousandth of an inch)

MOS CAPACITOR

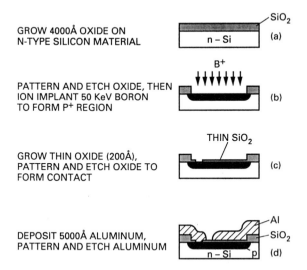

GROW 4000Å OXIDE ON
N-TYPE SILICON MATERIAL

SiO$_2$

n – Si

(a)

PATTERN AND ETCH OXIDE, THEN
ION IMPLANT 50 KeV BORON
TO FORM P$^+$ REGION

B$^+$

(b)

GROW THIN OXIDE (200Å),
PATTERN AND ETCH OXIDE TO
FORM CONTACT

THIN SiO$_2$

(c)

DEPOSIT 5000Å ALUMINUM,
PATTERN AND ETCH ALUMINUM

Al

SiO$_2$

n – Si   p

(d)

Processes (from top to bottom of figure) required to
fabricate a simple metal-oxide-semiconductor capacitor.

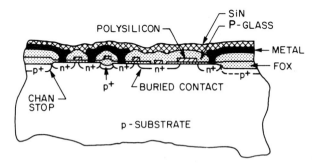

POLYSILICON

SiN
P-GLASS

METAL

FOX

p$^+$ n$^+$ n$^+$ n$^+$ n$^+$ n$^+$ p$^+$

p$^+$ BURIED CONTACT

CHAN
STOP

p - SUBSTRATE

Processes required to fabricate an n-channel metal-oxide-
semiconductor integrated logic circuit.

are involved, the precisions needed should be obvious. That is why clean rooms, white smocks, booties, and caps are so ubiquitous in the semiconductor business. A speck of dust or a hair are bulls in the china shop when working at these dimensions.

So far, we have been concerned with the manufacture of parts. Assembly, such as that done in the automobile plant that impressed my wife, is another critical part of production. At the time of the Industrial Revolution, products were assembled individually, in the same way that you might assemble a model airplane from a kit. This form of assembly is still quite common. Certainly the construction industry tends to build bridges, buildings, and freeways one at a time. When very expensive and complex items such as Cray computers and Boeing 747s are being assembled, one typically sees several in a room (a very large room in the case of 747s) in various stages of completion, with machines and workers clustered around them. Although several products are being built at once, and parts are made in multiples, in a sense each is being built individually.

Assembly is also often done in batch mode. A batch of products is made at one time, either on a highly flexible production line or by people working at stations that provide them with all necessary components for their task.

Perhaps the most impressive form of assembly is mass production. The products move along the line, which is designed so that the workers each have a particular job to do. Obviously, each of their jobs must take the same amount of time and be at least tolerable by a human for the duration of a shift. The design of an assembly line that not only does not exhaust or alienate the employees but keeps them satisfied and motivated to do good work is itself an extremely challenging engineering problem.

The great move toward mass production during the first half of this century involved designing an assembly line that would use not only people very efficiently but also specially built automatic machines. One still sees this in lines producing automobile engines, for example, where specialized machines completely assemble engine heads by putting the many parts (valves, springs, keepers, tappets, camshafts, bearings, nuts, bolts) on the machined heads in the proper location and order and tightening them to the proper specifications. A great amount of money and time is necessary to design, develop, and build such a specialized line, and once built, it only likes to assemble the type of engine head it was designed for.

An assembly line that can assemble a family of products, or even many different products, may represent less of an expense and gives the manufacturer the option of producing batches of products at will. This allows a smaller inventory and faster response to customers, all advantageous in present-day tight competition. One reason for the interest in industrial manipulators or robots, such as the welders shown earlier, is this flexibility. Their movements are controlled by computer and therefore can be easily reprogrammed to do a large variety of jobs.

Obviously we do not begin to cover the topic of production by talking about the manufacturing and assembly of hardware. In 1987 construction accounted for as much of the United States' GNP as the manufacture of machinery, electric and electronic equipment, and motor vehicles and equipment combined. And how about the so-called process industries—those who bring us our gasoline, beer, steel, plastic raw material, and aspirin tablets and furnish other industries with all of the many products they need? Process industries have different characteristics than industries dedicated to the manufacture and assembly of hardware. They require extremely careful control of temperatures, pressures, and chemical reactions, and deal with continuous flows of liquids and gasses. They are the domain of the chemical engineer, although one finds all disciplines in these industries.

Since the flourishing of the chemical industries during the Industrial Revolution, the process industries have benefited not only from increasing scientific knowledge but also from rapidly advancing technology. This has allowed improved control of processes, better and lower-cost materials, the simulation and design of increasingly complex and sophisticated systems and plants, and the use of catalysts, enzymes, and now genetically engineered microbes to aid in reactions. The invention of new classes of products such as plastics, fertilizers, pesticides, and pharmaceuticals has made major differences in our lives. The process industries now form a significant portion of our industrial capability. Some sense of this can be gained from the 1987 breakdown of the U.S. gross national product in Chapter 2.

If we look at all forms of production—manufacture and assembly of hardware and software, production of chemical products and raw materials, and construction—we find that many hardware components of the various processes involved were developed in their basic forms during the Industrial Revolution. Since 1850, however, production has been affected by the invention of new tools, such as the laser, the computer, and the bulldozer; by the use of new materials, such as epoxy; and by new insights into phenomena such as catalysis and kinetics. The designs of machine structures, mechanisms, and tools have become more sophisticated, and tremendous changes in the economics of production have become available because of economies of scale.

One of the major changes in production has come through developments in the *control* of the machines of production, especially with the advent of the digital computer. Initially, machines were controlled directly by human operators. Tools were often mechanically guided, but decisions as to how fast and in what direction the tools were moved were made by the operator, and the movement was started and stopped by the operator. In the nineteenth century, an operator "ran" each machine. Such manufacturing has not disappeared. Many modern machine tools intended for specialty work and low-production are of this type. The operators tend to be highly skilled and responsible for the quality of the final part.

The exceptions to this prior to this century were few and far between. One was the Jaquard loom, which still is used for weaving things such as carpet and the label on the inside of shirt collars. The position of the threads of the fabric is controlled by a number of "cards," which are large perforated plates that are secured to a continuous chain. These cards each have information on them that is coded by holes in the cards. As the cards move into position, they are read by a bed of wires which push against them. Wires passing through holes give the input to the loom as to the positioning of the proper thread. This loom is a precursor to the modern digitally controlled machine. Mechanical control was also sometimes obtained by use of "masters" and followers. Lathes used in the furniture industry, for instance, may be of this type. The desired shape of the product is contained in a metal template. The tool is guided mechanically by use of this template in order to ensure that the final product reaches the desired shape.

The next step was the application of electrical and hydraulic (pressurized fluid or gas) control to machines and the use of increasingly sophisticated sensors to provide feedback for "automatic" control. Electrical and hydraulic energy not only can be used to provide easily variable speeds and forces and to initiate and stop movements automatically but to allow information about the state of the desired output or intermediate stages in a process to be transmitted accurately and easily. The operator can now choose the desired level of output (position, velocity, acceleration, force, temperature, pressure, velocity, volume, or whatever). Sensors measure the actual output or pertinent intermediate variables in the process; they are compared with the desired ones, and the difference is used to make a correction in the system. This is the way that the cruise control in cars and the thermostat in houses work. The operator makes the initial decision, and the machine carries out the actual implementation. The process can be very complex, since the control hardware and software can be designed to take account of its behavior and act accordingly. Obviously, the applications of electrical and hydraulic control systems made a huge difference in production capability. Modern refineries could simply not be operated if each function had to be controlled by a human watching a sensor and turning a valve.

Digital circuitry has taken control even further. The digital computer allows us to process vastly increased amounts of information to a higher degree of accuracy for much less money. The computer is capable of handling complex interactions and therefore of making decisions on its own that would have previously required human intervention. Sometimes when the human does intervene, it turns out that the computer was right (as at Three Mile Island).

An interesting example of a product produced by benefit of a large amount of computer activity is the replica of the jawbones of a living human shown on page 196. It is the result of a process invented by Cemax, Inc. The input data are from either a CT or MRI scan. As described in Chapter 2, this data is a three-dimensional array of information and could not itself be generated without the computer. The Cemax process, first of all, uses a computer to find the particular data that define the surface of a particular body part. The computer then converts the data into a three-dimensional plot of the surface. It then converts this to a format usable by a numerically controlled milling

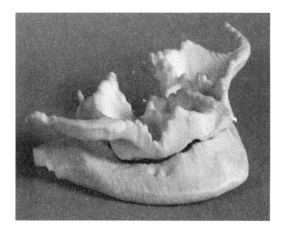

Jawbone models machined from Scan Data by CEMAX, Inc.

machine, which then follows the computer's orders to replicate the part. These parts are then used by surgeons, dentists, and other medical practitioners for implants or for designing procedures. For example, if someone fractures the facial bones on one side of his head and they do not heal evenly, the computer can subtract the image of one side of the patient's skull from the other and make a perfect implant from the difference. If you severely shattered your pelvis, wouldn't you like your surgeon to play around with replicas of the pieces before he sliced you open? The jawbones shown here were used to build implants to secure dental plates.

In addition to changes in the materials, processes, and machines used in production, there have been major changes in the format in which they are used. The "factory," whether it produces blenders, buildings, or benzene, has changed. These changes are more noticeable in industries that manufacture hardware products and in the construction industry, since severe social and economic forces have been brought to bear on these industries. Production in the chemical industry tends to be more dependent on expensive plant and less on human workers. Its problems therefore have had more to do with plant design and construction than the organization of large numbers of people, but it has certainly not been immune to this change. Bhopal certainly received public attention.

In the armor factory discussed earlier, the various parts of the armor being manufactured were similar but not especially "standardized." Production of the parts and the final product were based on hand labor and water and wind-powered hammers and polishing devices. Although workers were obviously expected to work "hard" and were trained through apprenticeship—we suspect that hours were long and that children were not exempt from the labor pool —nevertheless the factory does not include any time clocks, time-and-motion-study people, nonworking foremen, or schedules on the wall.

Much had changed by the time of the Ford Model T. This automobile owed its tremendous success to its usefulness, reliability, ruggedness, and low price. The low price was possible because of the factory built to produce the Model T, which was a marvel of standardization. It was the consummate mass assembly plant, with people performing carefully defined routine jobs repeatedly day after day. It did include time clocks, time-and-motion-study people, nonworking foremen, and schedules on the wall, but it was very successful. Its success was due not only to the excellent product and the large market for it but to a plentiful supply of competent immigrant labor with low expectations. In some ways, assembly line labor compares favorably with agricultural labor, and people who are new to a society are often willing to make sacrifices in order to become established and provide for their families.

Ford's attitude toward his employees would not be considered enlightened by present-day standards. However, it was typical of attitudes in the early years of this century, when the main interest of management was output and profit—to produce more with a given amount of labor. As a result of this mind-set, a whole field centered on the productivity of human workers evolved. Industrial engineering was being born. As factories became more complex, it became evident that the integration of people and machines was important. A much maligned pioneer in this field was Frederick Taylor, who along with colleagues such as Luther Gulick, Lyndall Urwick, Henry Fayol, and others developed the field of "scientific management." This included concepts such as hierarchical management with unity of command and limited span of control. In the early 1900s the models for successful organizations were the military, the church, and large governments, so that scientific management also assumed top-down authority, standardization of job design, and uniformity of behavior. Scientific management viewed people in factories as components of the overall machine and sought to design their jobs accordingly. At the time, the concepts of scientific management were extremely successful in increasing productivity, although now they seem rather inhumane. A famous quote from the period was a comment by Taylor about an ironworker named Schmidt. Taylor had redesigned Schmidt's job to increase his output (which was loading pig iron) by some 350 percent. Scientific management subscribed to the incentive system of payment, so for this increase in productivity Schmidt's salary was increased 60 percent, which pleased Schmidt a great deal and provided a happy ending to the story in these pre-union days. Taylor later made the following comment: "One of the very first requirements for a man who is fit to handle pig iron as a regular occupation is that he shall be so stupid and so phlegmatic that he more nearly resembles an ox than any other type." To the advocates of scientific management, the design of the job was critical; it should allow the worker to perform it repetitively, automatically, and incessantly. These days we would prefer the automaton to the ox, and in fact industry is strangely attracted to the robot. That is because present-day workers are not as phlegmatic as oxen.

One of the main tools of the early industrial engineers was the stop watch, and from this came the standardization that is endemic to the classic production line. Early industrial engineers are sometimes accused of having treated people like machines, and, in fact, in retrospect they are guilty. But at the turn of the century, this approach was extremely successful in increasing the productivity of manufacture.

Although many aspects of scientific management can still be seen in manufacture, attitudes toward workers have become much more enlightened, due to increased educational levels, unions, and a more cynical labor pool. In the 1930s the "human relations" movement was launched in industry, following the discovery that people's performance on the job is affected by psychological factors as well as physical ones. One well-known set of experiments that supported this concept occurred at the Hawthorne works of the Western Electric Company (the "Hawthorne" experiments, basis of the "Hawthorne" effect, which is the tendency of experimental subjects to perform better when they know they are being watched).

These experiments were to investigate the effects of illumination on productivity. Lighting levels were varied and productivity was measured in a test group of workers and in a control group that worked under standard illumination. Initially, illumination was increased and productivity began to increase in the test group, a finding that was consistent with the expectation of the experimenters. However, it also increased in the control group. The experimenters then began to decrease the illumination in the test group area and productivity continued to rise. At the conclusion of these experiments some of the workers were maintaining record productivity with an illumination level approximating full moonlight. Obviously something else was happening.

In another set of experiments, a special test room was set up in which a group of workers assembled products. In order to control all experimental variables, the workers were given periodic medical checkups, and full records were kept on their sleep and other personal activities. A full-time observer in the room kept an accurate journal of all that happened. The workers were also interviewed periodically, and during the interview they were offered ice cream and cake. Over two years, working conditions were varied widely. Yet productivity followed a general upward trend.

At the time of these experiments, the conclusion that people produced more if they feel noticed and if they are humanely managed was revolutionary. Psychological factors simply had not been considered by the principles of scientific management, and their study was obviously something that could benefit industry. The human relations movement focused on raising the morale and motivation of workers. Effort was spent improving the physical environment and the nature of the interaction between bosses and workers; companies hired psychologists and began to view themselves more as collections of individuals than as structures with replaceable human components. Top-down management and detailed job descriptions were still in place, but work became more pleasant. Stop watches still existed, but unions were now constraining their use.

During World War II tremendous demands were put on industry, and because many younger men were in the military, the industrial labor pool saw an influx of nontraditional workers, such as women. Human engineering approaches allowed work stations, tasks, and tools to be designed to fit the workers who would use them. Motivation, commitment, and cooperation were high during the war, and the technical sophistication of products and processes (such as the application of large-scale welding to produce Liberty ships in very short time intervals) was growing. These factors all contributed to the United States' emergence from World War II as the internationally dominant power in production.

Rising income and expectations coupled with union activities soon led to even more concern with the quality of life of the production worker. Salaries and benefits were increased, workweeks were shortened, cafeterias, restaurants, and workspaces were improved, noise levels were lowered, and safety was emphasized. Automation replaced much tedious manual labor, although it also made obsolete many highly skilled and therefore satisfying jobs. The individual worker was no longer a wage slave, as he was in the days of the Model T. There was still concern over the effects of mass production on workers, and assembly line jobs did not rank high on the desirability list, but wages were adequate to fill the jobs, and the productivity of the U.S. labor force continued to rise.

The design of production plants and processes also became more sophisticated after World War II. Specialists in the fields of operations research and production management developed powerful new scheduling tools and techniques of optimally locating plants and designing assembly lines better so that each worker would have a task that could be accomplished well in the available time. Better approaches to keeping track of and controlling inventory were developed, and the use of statistical quality control became routine. The digital computer was not only able to keep track of such data, but could convert them to forms easily used by management.

However, as other countries have recovered from World War II, and as the United States has found itself struggling harder to compete with their modern plants and high motivation, our philosophy and approaches to production have changed. This has been partly because of the realization that production is key in this competition, but it is also due to new technology and changed social attitudes.

We have learned, for instance, that in certain instances we must give production workers more complex tasks to relieve boredom. We learned this

partly through experiments in Europe, such as Volvo's innovative production plant at Kalmar, Sweden, in which teams of workers are assigned the task of assembling an entire automobile (back to the armor factory, only with standardized parts). We have learned that workers understand their work better than managers do. The Packard Division of General Motors built an extremely innovative plant in Mississippi that was entirely run by hourly workers. Their mission was to build cable-harness assemblies to certain technical and economic specifications. They worked in teams, subcontracted work to cottage industries, adopted flexible assignments and schedules, and met all of their goals. Many U.S. companies have adopted quality circles and other mechanisms to allow closer interaction between workers and managers, with resulting success. We have learned that we should give workers more responsibility for improving the way their task is done and the quality of the final product.

We have learned that inventory represents investment, and have adopted "Just-In-Time" approaches that bring material into the factory as it is needed. This has required better relationships between factories and the many suppliers who provide them with materials and components. We have learned to increase the quality of our products many orders of magnitude by changing attitudes and expectations of both managers and workers, by paying more attention to products in all phases of engineering, by becoming more sophisticated at measurement, and by better integrating suppliers of materials and components into the production process so that they will assume more responsibility for the quality of their products.

We are now working on decreasing the time it takes new products to reach the market, both to decrease costs and to respond better to changes in market demand. One advantage of the present move to improve the United States' production capability is that we seem to have finally become willing to learn from others, in particular Japan. We are studying their "Kanban" approach to manufacture and learning about flexible assembly lines that can very rapidly be changed from one product to another.

An example that has received quite a bit of attention is a line designed by Allen-Bradley in Milwaukee[2] to produce four styles of contactors (motor switches) and three styles of electrical relays that can have up to 1,025 different customer specifications. This line can produce these products in lot sizes down to one within twenty-four hours of the order much more cheaply than the previous traditional line. For instance, the cost of a typical contactor was reduced from $13.25 to $8.35. Since only 10 percent of the old cost was direct labor (reduced to zero in the new line), this lower price represents an appreciable savings in overhead. This savings was accomplished by eliminating the need for warehousing, inventory, supervisory people, and inspectors. At this new cost the products are easily globally competitive.

Orders are entered into computers at the field sales office of Allen-Bradley and then transferred to the company main-frame computer in Milwaukee. Each morning at 5:00 a.m. the previous day's orders are downloaded to the smaller computer that controls the line. This computer translates the orders into detailed production requirements and sends the appropriate information into a master controller that is manufactured by Allen-Bradley. This controller then gives inputs to the 26 controllers that tell the machines on the factory floor what to do. The line can run at different speeds, depending on the workload, and can produce 700 completed units per hour. Parts are produced automatically as needed with a small enough backlog that it can be stored on-line. Statistical quality control is incorporated into the production process, which includes more than 3,500 data collection points and 350 assembly test points to check the product as it moves along the line. After assembly, the units are automatically packaged, labeled, and sorted into the proper orders so that they can be immediately shipped. There are no people actually involved in the manufacturing and assembly (over 100 people would be needed to assemble the products manually). Seven people oversee the line, maintaining machines when necessary (lights go on when they are having problems) and ensuring that the raw materials are available. The company is experimenting with automating much of the handling of raw materials, and with a computer-based voice-simulation system that would tell the human operator if anything was wrong on the line. Allen-Bradley hopes to be able eventually to run the line with a single person.

We are also, finally, reintroducing manufacturing and production into our educational system. Engineering and business schools are paying more attention to operations management, manufacturing, and other aspects of production. Joint industry–university programs and new degree programs are being developed in a number of locations, often with support from newly sensitized government agencies. Competitiveness has become a buzzword in the United States, and much of the emphasis is on production.

We will now turn to the relationship of engineering and money—the intimate tie between technology and business.

9

Money and Business

The Grease

By now it should be clear that engineering is very seldom a solo activity. Even Hero of Alexandria probably scrounged a bit of monetary support from the priests before he built their automatic temple-door openers, and no doubt he solicited help from local technicians in manufacturing some of the parts. These days much of technology involves large amounts of money and lots of people. The engineer must work in this environment and is constrained by the availability of these resources.

This is slightly ironic, because according to the stereotype, engineers like "things" better than people and are not particularly financially oriented—they are content to sit at their workstations or work in their laboratories and be brilliant. Engineering students are often taken in by this stereotype and have chosen engineering because they think it's a way to avoid worrying about office politics and the "bottom line." When I encounter such students, I chuckle quietly to myself, knowing how quickly they will discover the opposite. In fact, successful engineers usually become good managers of both people and money.

Typically, the newly graduated engineer begins work as an informal apprentice. This training period is necessary because many aspects of engineering are not taught in school. Engineering students who are about to graduate often have an uncomfortable feeling that they have not learned enough practical knowledge and technique to be engineers, and they are right. But this lack of knowledge is not a long-term problem, nor is it a surprise to their employers. Schools do not need to teach things that can be better learned on the job. Not only are such things job-specific, but they are very difficult to teach without the motivational setting available in industry. Knowing when and how to get help, for instance, is critical in the engineering role, but it is not emphasized in school; schools are designed to evaluate the work of individuals, and they stress competition rather than cooperation. Because time is short, and so much ground must be covered by every student, most schools give their students little time to work on group projects, and on saving a few cents here and there on assembly operations, such as putting a wheel on a car. There is no way to convey the importance of this in a school. One has to be working for an automobile company, be part of the team, and see the result of such a saving spread over millions of cars.

The period of apprenticeship varies in length, depending on how closely one's job matches one's education. In work that requires modern theory (mechanics, software design, solid-state circuit design), the apprenticeship may be quite short, since graduates often arrive at the job with knowledge that is more current than that of the senior engineers. In more pragmatic fields (manufacturing, machine design, oil production) it may take years to learn the ropes.

However, even in this apprenticeship role the young engineer will be expected to stay within budget and to interact with other engineers, office staff, clients, purchasing and shop people, assembly workers, and whoever else is necessary to get the job done. After a bit of time (a year or so) the young engineer will find that even younger engineers have appeared on the job, and they will need advice and help in turn. The role of mentor comes fast, which is why very young people in industry may bear the title "senior engineer."

The next step is formal responsibility over a group or small project. This can happen quite soon. When I joined the Air Force six months out of college, I became an officer and therefore had responsibility for projects, even though I didn't know what I was doing. At the Jet Propulsion Lab I was given instant responsibility for designing pieces of hardware; after two years I led a group of engineers working on future spacecraft designs; after three years I was "promoted" to group supervisor, in charge of the work of 15 people or so.

As soon as one moves into formal management, one cannot avoid dealing with budgets, schedules, and employees' neuroses, motivations, salaries, personality clashes, love affairs, illnesses, ambitions, and values in addition to technical matters. Before long, one is also likely to be involved in setting budgets, defining future projects, and shaping company priorities. As one acquires increasing responsibility, one eventually becomes more involved in general management than detailed technical work. Such is the fate in store for many engineering students. Many of them will say, while in school, that they will not accept managerial work. However, few of them will decline it when offered. Management jobs allow one to participate in a bigger piece of the picture. They usually offer more income and appear to offer more prestige, influence, and stimulation than straight technical work.

The reason that engineers become managers is twofold. First of all, many traits of engineers are valuable in management. Engineers are analytical, they approach problems quantitatively, they typically work hard, and they have confidence in their ability to "fix" things. For these reasons they are admitted to graduate schools of business in large numbers. Second, managers in technology-based companies, although they must deal with people and money like all managers, must also understand the technology if they are to make intelligent decisions about project and research directions. They must be technically competent to retain the respect of the engineers who report to them. Organizations have found it easier to take managers from the engineering ranks than to teach engineering to nontechnical general managers.

This behavior leads to a conflict in many technology-based companies. Top managers say that they would like their better engineers to remain in detailed technical work, since they see it as invaluable to the success of their companies. Some companies have even installed a so-called dual ladder system that attempts to provide the same rewards to both technical and managerial people for high achievement and seniority. When engineers are offered management jobs, however, the lure of involvement in decisions at a company-policy level and of getting one's fingers in more pies is difficult to withstand. In addition, though many companies say they want their engineers to remain in technical activities, they seem to give bigger paychecks, larger offices, better parking spaces, and more respect to those engineers who become managers.

Let's go into a bit more detail first about money. How much money does an engineer cost? If we look at the United States, we find about 2 million engineers and a gross national product of about $4.25 trillion. Of this GNP, approximately $1.5 trillion is in the areas of construction, manufacturing, transportation, public utilities, and communication, all of which would appear to be somewhat based on engineering. Let's throw in another $0.5 trillion to cover other areas in which engineers work (it will also make the arithmetic easier). All of the money from these sectors is not spent for or by engineers. Remembering our quantitative thinking exercises, we might guess that 20 percent of it is, and conclude that each engineer in the United States spends $200,000 per year. The 1988 edition of the *Statistical Abstract of the United States* lists a cost per scientist or engineer involved in research and development in private industry of $140,000 for 1985. Given inflation and the fact that some engineers cost less, some more, perhaps we are in the right ballpark. Looking at this problem in a different way, we know that some organizations are primarily engineering organizations. The Jet Propulsion Laboratory does not produce consumer products nor make a profit; it is in business simply to design, build, and operate new devices. It employs about 4,300 engineers, scientists, and technicians and has an annual budget of about $1 billion. Dividing these figures gives us about $230,000 per technical person—a higher number, but the spacecraft business is expensive. Actually, the more direct cost of an engineer at JPL (salary, benefits, facility needs, utilities, and office space) averages only about $110,000; the $120,000 difference is the cost of implementing the thinking that they do. Organizations such as JPL have many essential employees who are not technical (about 1,300 in JPL's case), but in a sense they support the engineering operation and would therefore be included in the $120,000.

As an example at the other end of the spectrum, a friend of mine who is a geotechnical consultant tells me that it costs him about $100,000 to maintain an engineer in his office. Since consulting companies do not build hardware or produce software, this figure is simply the cost of space, utilities, salary, benefits, travel, telephone, computer time, and office expenses.

Engineers are big users of computers and also require the support of people such as technicians and draftsmen. These things cost money. The design process requires the construction of models and prototypes, and such one-of-a-kind items are extremely expensive. Engineers speak of the learning curve, which represents the cost of a product over time. Initially the cost of a product is extremely high. As time passes, people become familiar with the process, and new innovations simplify production. Economies of scale occur as supplies and equipment are purchased in larger quantities. Production costs decrease as plant and equipment is amortized over time. Prototypes are not even on the curve. For instance, it costs half a million dollars to produce a prototype automobile, since all parts have to be fashioned by hand.

Engineers produce ideas that may be patentable, and, as we have seen, money is needed for the process of filing for and defending patents. Finally, they produce designs that are manufactured, sold, and maintained. Production

OUR
$18.00 Giant Power Heidelberg Electric Belt

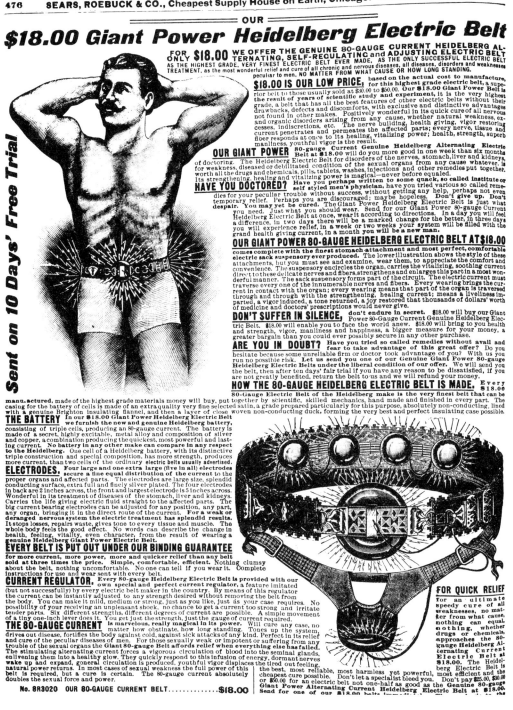

Sent on 10 Days' Free Trial

FOR ONLY $18.00 WE OFFER THE GENUINE 80-GAUGE CURRENT HEIDELBERG ALTERNATING, SELF-REGULATING and ADJUSTING ELECTRIC BELT AS THE HIGHEST GRADE, VERY FINEST ELECTRIC BELT EVER MADE, AS THE ONLY SUCCESSFUL ELECTRIC BELT TREATMENT, as the most wonderful relief and cure of all chronic and nervous diseases, all diseases, disorders and weaknesses peculiar to men, NO MATTER FROM WHAT CAUSE OR HOW LONG STANDING.

$18.00 IS OUR LOW PRICE, based on the actual cost to manufacture, for this highest grade electric belt, a superior belt to those usually sold at $30.00 to $50.00. Our $18.00 Giant Power Belt is the result of years of scientific study and experiment, it is the very highest grade, a belt that has all the best features of other electric belts without their drawbacks, defects and discomforts, with exclusive and distinctive advantages not found in other makes. Positively wonderful in its quick cure of all nervous and organic disorders arising from any cause, whether natural weakness, excesses, indiscretions, etc. The nerve building, health giving, vigor restoring current penetrates and permeates the affected parts; every nerve, tissue and fiber responds at once to its healing, vitalizing power; health, strength, superb manliness, youthful vigor is the result.

OUR GIANT POWER 80-gauge Current Genuine Heidelberg Alternating Electric Belt at $18.00 will do you more good in one week than six months of doctoring. The Heidelberg Electric Belt for disorders of the nerves, stomach, liver and kidneys, for weakness, diseased or debilitated condition of the sexual organs from any cause whatever, is worth all the drugs and chemicals, pills, tablets, washes, injections and other remedies put together. Its strengthening, healing and vitalizing power is magical—never before equaled.

HAVE YOU DOCTORED? Have you perhaps written to some quack, so called institute or self styled men's physician, have you tried various so called remedies for your peculiar trouble without success, without getting any help, perhaps not even temporary relief. Perhaps you are discouraged; maybe hopeless. Don't give up. Don't despair. You may yet be cured. The Giant Power Heidelberg Electric Belt is just what you need. Just what you should wear. Send for our Giant Power 80-gauge Current Heidelberg Electric Belt at once, wear it according to directions. In a day you will feel a difference, in two days there will be a marked change for the better, in three days you will experience relief, in a week or two weeks your system will be filled with the grand health giving current, in a month you will be a new man.

OUR GIANT POWER 80-GAUGE HEIDELBERG ELECTRIC BELT AT $18.00 comes complete with the finest stomach attachment and most perfect, comfortable electric sack suspensory ever produced. The lower illustration shows the style of these attachments, but you must see and examine, wear them, to appreciate the comfort and convenience. The suspensory encircles the organ, carries the vitalizing, soothing current direct to these delicate nerves and fibers, strengthens and enlarges this part in a most wonderful manner. The sack suspensory forms part of the circuit. The electric current must traverse every one of the innumerable nerves and fibers. Every wearing brings the current in contact with the organ; every wearing means that part of the organ is traversed through and through with the strengthening, healing current; means a liveliness imparted, a vigor induced, a tone returned, a joy restored that thousands of dollars' worth of medicine and doctors' prescriptions would never give.

DON'T SUFFER IN SILENCE, don't endure in secret. $18.00 will buy our Giant Power 80-Gauge Current Genuine Heidelberg Electric Belt. $18.00 will enable you to face the world anew. $18.00 will bring to you health and strength, vigor, manliness and happiness, a bigger measure for your money, a greater bargain than you could ever possibly secure in any other purchase.

ARE YOU IN DOUBT? Have you tried so called remedies without avail and fear to take advantage of this great offer? Do you hesitate because some unreliable firm or doctor took advantage of you? With us you run no possible risk. Let us send you one of our Genuine Giant Power 80-gauge Heidelberg Electric Belts under the liberal condition of our offer. We will send you the belt, then after ten days' fair trial if you have any reason to be dissatisfied, if you are not greatly benefited, return the belt to us and we will refund your money.

HOW THE 80-GAUGE HEIDELBERG ELECTRIC BELT IS MADE. Every $18.00 80-Gauge Electric Belt of the Heidelberg make is the very finest belt that can be manufactured, made of the highest grade materials money will buy, put together by scientific, skilled mechanics, hand made and finished in every part. The casing for the battery of cells is made of an extra quality very fine selected satin, a grade prepared particularly for this purpose, absolutely non-conducting, lined with a genuine Brighton insulating flannel, and then a layer of close woven non-conducting duck, forming the very best and perfect insulating case possible.

THE BATTERY In our $18.00 Giant Power Heidelberg Electric Belt we furnish the new and genuine Heidelberg battery, consisting of triple cells, producing an 80-gauge current. The battery is made of a secret, highly excitable, metal alloy and composition of silver and copper, a combination producing the quickest, most powerful and lasting current. No battery in any other make can compare in any respect to the Heidelberg. One cell of a Heidelberg battery, with its distinctive triple construction and special composition, has more strength, produces more current, than two cells of the ordinary electric belts usually advertised.

ELECTRODES. Four large and one extra large (five in all) electrodes secure a fine equal distribution of the current to the proper organs and affected parts. The electrodes are large size, splendid conducting surface, extra full and finely silver plated. The four electrodes in back are 2 inches across, the front and largest electrode is 5 inches across. Wonderful in its treatment of diseases of the stomach, liver and kidneys. Carries the life giving electric fluid straight to the affected parts. The big current bearing electrodes can be adjusted for any position, any part, any organ, bringing it in the direct route of the current. For a weak or deranged nervous system the electric treatment has splendid results. It stops losses, repairs waste, gives tone to every tissue and muscle. The whole body feels the good effect. No words can describe the change in health, feeling, vitality, even character, from the result of wearing a genuine Heidelberg Giant Power Electric Belt.

EVERY BELT IS PUT OUT UNDER OUR BINDING GUARANTEE for more current, more power, more and quicker relief than any belt sold at three times the price. Simple, comfortable, efficient. Nothing clumsy about the belt, nothing uncomfortable. No one can tell if you wear it. Complete instructions for use and wear sent with every belt.

CURRENT REGULATOR. Every 80-gauge Heidelberg Electric Belt is provided with our own special and perfect current regulator, a feature imitated (but not successfully) by every electric belt maker in the country. By means of this regulator the current can be instantly adjusted to any strength desired without removing the belt from the body. You can make it mild, medium or strong, just as you like, just as your case requires. No possibility of your receiving an unpleasant shock, no chance to get a current too strong and irritate tender parts. Six different strengths, different degrees of current are possible. A simple movement of a tiny one-inch lever does it. You get just the strength, just the gauge of current required.

THE 80-GAUGE CURRENT is marvelous, really magical in its power. Will cure any case, no matter how obstinate, how long standing. Tones up the system, drives out disease, fortifies the body against cold, against sick attacks of any kind. Perfect in its relief and cure of the peculiar diseases of men. For those sexually weak or impotent or suffering from any trouble of the sexual organs the Giant 80-gauge Belt affords relief when everything else has failed. The stimulating alternating current forces a vigorous circulation of blood into the seminal glands, enlivening them into a healthy glow. They quickly respond to this infusion of energy, dormant nerves wake up and expand, general circulation is produced, youthful vigor displaces the tired out feeling, natural power returns. In most cases of sexual weakness the full power of this belt is required, but a cure is certain. The 80-gauge current absolutely doubles the sexual force and power.

No. 8R3020   OUR 80-GAUGE CURRENT BELT..............$18.00

FOR QUICK RELIEF for an ultimate speedy cure of all weaknesses, no matter from what cause, nothing can equal, nothing, whether drugs or chemicals, approaches the 80-gauge Heidelberg Alternating Current Electric Belt at $18.00. The Heidelberg Electric Belt is the best, most reliable, most harmless yet powerful, most efficient and the cheapest cure possible. Don't let a specialist bleed you. Don't pay $25.00, $30.00 or $50.00 for an electric belt not one-half as good as the Genuine 80-gauge Giant Power Alternating Current Heidelberg Electric Belt at $18.00. Send for one of our $18.00 belts...

requires money for building plants, purchasing parts and materials, installing machines, tooling, and of course paying people. Marketing, distribution, and maintenance of products require more people and the money to support them. In other words, supporting engineers and implementing the ideas they produce require a huge outlay of money. But without those ideas the company's income would be zero.

Engineers play an equally active role in making money. In the steady state, it comes from the sale of products and services to individuals, other companies, or governments. The U.S. government is a major supporter of engineering activities. In addition to funding approximately one half of all research and development in the United States, the government also purchases just about anything you can imagine, from astroturf to body bags, and the production of all of these products requires engineers also. For instance, in 1986 the amount of federal funding for defense R&D (in the Departments of Defense and Energy) was $36.9 billion, but the total outlay for defense was $273.4 billion, of which $70.8 billion was for aircraft, $8.7 billion for ships, $12.2 billion for missiles, $4.0 billion for ammunition, and $6.8 billion for structures. Engineers are needed to build such things, even if the designs are not new.

The majority of the money spent by government goes to private industry, although some goes to government laboratories, some to nonprofit laboratories, and some to universities. Engineers play a role in securing this government money as well as in spending it. Government money for new developments is allocated in response to proposals. These proposals are usually in response to a published government need (a request for proposal), although they can be unsolicited. They not only include an overall description of the proposed product or service, with budgets and schedules, but often detailed

A product proudly advertised in a 1902 edition of the Sears Roebuck catalogue, illustrating the appeal of technology to U.S. society at the turn of the century. The catalogue is filled with products of industry, many of which are intended to improve the customer's technical ability. The mail-order business was and is a very successful way of distributing such products.

feasibility studies that require a significant amount of engineering input. I spent quite a bit of time at JPL on studies that would later appear in proposals to NASA for new missions. I have often been called upon to "sell" such proposals at presentations and briefings. This is typical for engineers whose work is funded by government money.

If a government proposal is successful, it results in complex contractual relationships ranging from a fixed fee for simple predictable outputs to arrangements that combine cost reimbursement, fees for services, and various incentives—for early delivery, completing the work under the proposed budget figure, performance of prototypes, and so on. Engineers furnish input to the negotiation of such contracts also, because they are the ones who have a feeling for the uncertainties involved.

The largest source of money that supports engineering in the United States, however, is not the government; it is direct sales either to individual consumers, to distribution outlets and networks, or to other industries. Here too engineers are often found directly involved in the money aspects of the business. For instance, many engineers work directly in marketing and sales, especially if the products are technologically sophisticated. If one intends to sell gas chromatographs, nuclear reactors, or high-vacuum systems, one must not only understand the products in great detail but must also be able to speak the language of other engineers. Stanford engineering graduates are often surprised, when they go to work for a company such as IBM, to find that they are beginning their career in marketing and sales. Many successful technology-based companies not only believe that marketing and sales people should be engineers but that engineers should understand the customer and therefore have experience in marketing and sales.

Inattention to the customer was a weakness of many American technology-based companies in the 1970s. In fact, many consultants and authors became rich during the 1980s selling the message of becoming "closer to the customer." Much effort was spent in ensuring that engineering and marketing functions interacted more successfully and that engineers gained a better understanding of the end customer for their efforts. I am familiar with several companies who routinely require that their engineers spend time with customers, if not on sales or service calls, at least informally—once again, a task that may seem unusual to the young engineer, who may have expected a life confined to calculations and the laboratory.

Although much money comes eventually from sales of products and services, it is sometimes necessary or attractive to raise money from other sources—from bank loans, from sales of stock, from venture capitalists, or from limited partnerships for research and development, joint ventures, or, during the 1980's, junk bonds. Money from these sources is particularly needed for new operations or start-up companies, and, again, engineers are useful in raising this money, assuming that it requires selling a potential investor on the feasibility of a new technology. Engineers have been particularly active in starting companies in high-technology areas such as electronics and computers during the past thirty years.

In order to be healthy, a fledgling company in such an area has to have competence in finance, marketing, the management of people, *and* technology. Many major companies started small, and were the brainchild of engineers (the background of Messieurs Hewlett and Packard). Most start-ups in areas of sophisticated technology include engineers. They are necessary not only for the design and manufacture of the products or services but for the preparation of the business plan upon which the funding proposal is based. People who furnish capital to new companies are not naive. In fact, many venture capitalists have engineering backgrounds themselves and have held a wide variety of jobs in various companies. A believable business plan must be convincing not only in its technical feasibility and market potential but also in the details of its engineering functions such as research, design, development, and production.

Now let us turn from money to people. In what sort of human environment do engineers usually work, and what type of interactions are necessary? The majority of engineers work in formal organizations, whether big or small, that are hierarchical. However, organizations are not as simple as they appear in the charts. For instance, in the organizational chart of Hewlett-Packard there are several types of suborganization (see next page). The Measurement Systems Sector still retains an organizational style reminiscent of the company in its younger years. Somewhat independent divisions in various geographical locations have responsibility for product lines. These divisions are largely

HEWLETT-PACKARD CORPORATION ORGANIZATION

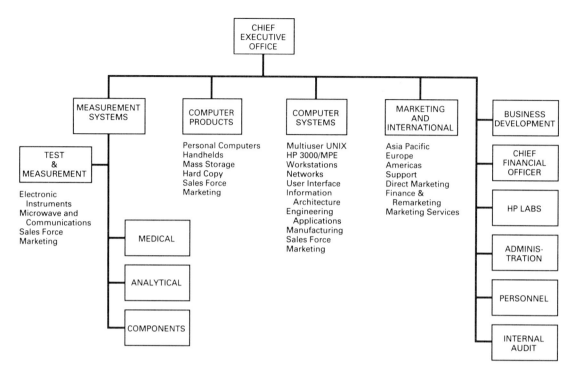

Hewlett-Packard organizational chart.

self-contained and carry out many of the functions of an independent small company. The Computer Products and Computer Systems Sectors, on the other hand, are heavily integrated. They are big companies in their own right. HP Labs and Marketing and International are corporate activities that must interact with people in related activities throughout the company. The remainder of the boxes are corporate administrative groups.

To further complicate the human environment, organizations are presently attempting to increase communication between people engaged in related functions. The present competitive economic environment does not allow the sluggishness, waste, and lack of creativity of a strictly top-down structure. A diagram originated by the Leaders for Manufacturing Program at the Massachusetts Institute of Technology gives an indication of this new approach (see figure opposite). The traditional neat and tidy hierarchical organizational chart does not indicate the amount of communication that must occur between

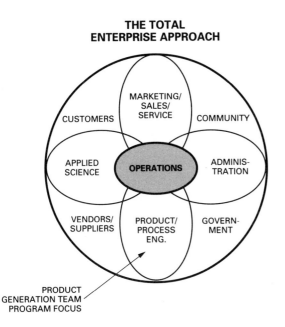

THE TOTAL ENTERPRISE APPROACH

CUSTOMERS

MARKETING/ SALES/ SERVICE

COMMUNITY

APPLIED SCIENCE

OPERATIONS

ADMINIS- TRATION

VENDORS/ SUPPLIERS

PRODUCT/ PROCESS ENG.

GOVERN- MENT

PRODUCT GENERATION TEAM PROGRAM FOCUS

Functional interactions in a manufacturing organization.

functions to compete in today's world. Technology-based organizations are trying to increase communication within engineering and between engineering and other related functions both inside and outside of the company.

Engineering functions may be distributed in different parts of the company, whether these parts are described by product, type of operation, or location. They may be centralized under a chief engineer or engineering vice president. Depending on the number of engineers, they may then be subdivided into divisions, sections, groups, laboratories, or other levels in the organization. It is typical to group engineers by technical specialty (the aeronautics section, the heat-transfer group). Although some engineers have a specialist role, acting as a consultant to others, most have an allegiance to the project they are working on at the time.

This is one of the schizophrenias of engineering life. One's expert knowledge and skills come from the community of engineers doing similar things.

If I am a software designer in a company, I will glean ideas, approaches, knowledge, techniques, and perhaps inspiration from other software designers. But when I am working on a specific project, a large percentage of my interactions must be with the other engineers working on that project, many of whom represent different technical disciplines. Project work requires the ability to be comfortable with interdisciplinary problems and to communicate in several specialized languages. Project work, because it involves more parameters, has a higher level of unknowns. Management in engineering must deal with both intellectual specialization and project work.

One approach to solving this management problem is a matrix organization, which assigns each person to both a discipline and a project. Such an approach may be extremely formal: Organizational charts may exist on paper showing each person's role in both discipline and project; one may have both a discipline and a project boss; and formal reviews may be based on performance both in the discipline and in the project. But the more typical case is less formal, in that people are assigned to either a discipline or a project and are known to be involved in the other. Managers in both disciplinary and project areas arrange formal and semiformal activities to ensure that both types of interactions occur. Larger organizations encourage ongoing disciplinary education and training and will sometimes sponsor conferences and meetings on disciplinary activities. They will also make sure that everyone is kept up to date on project activities.

Project managers in technology-based companies are typically engineers, since such work requires a high degree of technical understanding. The project manager's responsibility is to integrate the many disciplines and ensure that the project stays on schedule, within budget, and within specifications. Good project managers must be able to handle an immense amount of technical complexity and must be excellent leaders, since team members go through stages of frustration, panic, and fatigue as well as creativity and elation.

When managers evolve from an engineering discipline, one problem that often arises is a confusion between managerial duties and specialist work. Engineers tend to evaluate themselves foremost in terms of their technical competence. They are also intellectually competitive people, proud of their knowledge and ability to do their work well. This sometimes leads to a conflict of interest, as managers find themselves technically competing with the people who report to them, and this weakens their ability to manage fairly and professionally. In a classic case study originally written by A. Bavelas and G. Strauss, an engineer named Bob Knowlton is head of a group of engineers engaged in advanced development.[1] Bob's boss introduces him to a man who has just been hired by the company and asks Bob to tell the new man something about what is happening. It turns out that the new man is a whiz, likes Bob, and requests that he be in Bob's group. He proves to be very good at technical work but he does not particularly like to attend meetings and work with the other members of the group, and this attitude presents problems for Bob as a manager. At a meeting in which Bob makes a presentation, the new man, who has solved a very important problem, gets the majority of attention. Bob, feeling increasingly overshadowed, finally concludes that the new man is going to be given his job and so he resigns and takes a job with a competing company. Bob's former boss is dumbfounded, because the new man is not interested in management, Bob was a terrific manager, and now the group no longer has a leader. Bob's mistake was to judge his worth to the company solely by his own technical capability, rather than to realize that his job was to supervise a person whose technical skills were better than his. In fact, good technical managers look for people to hire who are technically better than they are.

Another task of managers in technical companies is to coordinate related functions which, if left alone, might diverge. An example is the necessary cooperation of product designers and marketing people. Designers think in terms of improving the product along functional and aesthetic lines. Marketing people look outward and think in terms of what people will buy. At some point they must come to some agreement about the product. Another example is the interaction between design and manufacturing people. If left alone, designers would tend toward more sophistication of form and function even if manufacturing became more difficult. Manufacturing people tend to value ease of production more highly. Once again, cooperation and consensus are necessary.

In technology-based companies, managers must be able to deal with a diversity of problem-solving styles. Some people are strong in mathematics, while others tend to think visually. Some people are visionaries, and some are nuts-and-bolts pragmatists. Some people approach problems analytically and intellectually, while others rely more on experience and intuition. The Myers Briggs Type Inventory,[2] a survey that measures preferences in problem-solving styles, spreads people across four axes, the extremes of which represent opposite styles. The adjacent figure shows how a group of managers from technology-based companies with engineering backgrounds fared according to this index. The axis at the top of the figure spreads people between those who prefer to implement ideas (E) and those who prefer to improve them (I). Those in the middle do not have a strong preference. The second ranges from those who prefer data, numbers, books, experts, and laboratories (S) to those who depend upon their intuition (N). The third discriminates between those who prefer to think their way to decisions (T) and those who would rather rely on feeling (F). The final one ranks people between those who are extremely judgmental (J) and those who adapt readily to new inputs (P).

Compared with the population at large, these particular managers seem to congregate at the I, S, T, and J poles. They are pathetically weak in representing that part of the populace that prefers to make decisions by feeling. But the spread is large, and this range is healthy for an organization, since all of these problem-solving strategies are necessary. However, people at the extremes do not always work together easily. Consider the J–P axis, for instance. It is very valuable to have both judgmental and adaptive people in an organization, since judgmental people cause the problem-solving process to converge, and adaptive people are comfortable with new inputs. But the words "waffler" and "pighead"—hardly terms of respect—are often used by people on the extremes of this axis to describe those on the other end in times of frustration. This makes problems for managers, especially if these people with different styles are more expert than the manager in sophisticated technical specialties. It may be that the difference between the aerodynamics expert who wants to build it and the one who wants to analyze it further is due more to a difference in problem-solving preference rather than to a difference of opinion about the merits and shortcomings of a particular wing shape.

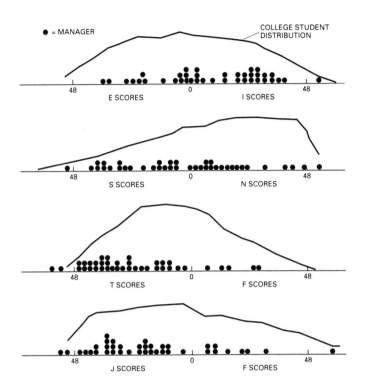

Myers Briggs Type Inventory of managers with engineering backgrounds (see text).

A good manager must also ensure that adequate communication is occurring between engineers and others in the company who do not have technical backgrounds. Here one finds the two-culture problem so nicely defined by C. P. Snow and still alive in universities (where students sometimes refer to themselves as techies and fuzzies) and in the larger society (I went to both engineering school and art school, but I always talk about art at parties). Engineering is loaded with specialized jargon and shorthand; it is thick with TLA's (Three Letter Acronyms). Nontechnical people do not like to be subjected to language they don't understand, and engineers are often frustrated by people who do not think as quantitively and scientifically as they do.

Managers must ensure that the proper communication is taking place between people with technical backgrounds and people without them in business and support functions who must cooperate on a given job.

Finally, in addition to managing money and people, engineers often wind up in positions that require them to help formulate and manage their company's long-term technology strategy.[3] In the short term, businesses employing engineers must produce and sell products that fit the market. However, in order to succeed in the long run, these organizations must worry about their technological capability. They must ensure that they have the knowledge and skills necessary to compete successfully. If they fall short in these areas, they will eventually fail in the marketplace. If they have too much capability, they may also fail because of the expense of supporting people and apparatus that are not employed on money-making projects.

Before and directly after World War II, firms in industries where technology was changing slowly, such as the automotive and steel industries, merely had to acquire new engineers to replace those who departed, and to provide for growth. Firms in industries where technology was changing more rapidly, by contrast, such as the electronics and communication industries, had to support in-house research and development efforts (such as Bell Labs) and remain current with pertinent developments in universities, government laboratories, and the professional literature. However, for most U.S. firms in this period business was simpler than at present. One reason for this was the small amount of interaction between firms in different industry sectors and between firms in different countries. Companies could succeed by paying attention to traditional capabilities in design, manufacturing, and R&D and did not have to consider technological capability as a central component of company strategy.

The situation has now changed dramatically. As we discussed in Chapter 8, manufacturing technology is changing rapidly and is now a key to success. Firms in fields such as genetic engineering, semiconductors, and computers consider their technology to be their main competitive asset and routinely engage in strategic partnerships with other companies in order to gain strength in functions such as manufacturing and marketing and to augment their technology. Foreign competitors, such as Japanese firms, obviously consider technology a strategic weapon. The VLSI (Very Large Scale Integration)

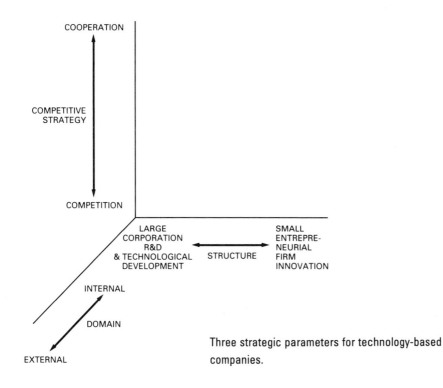

COOPERATION

COMPETITIVE
STRATEGY

COMPETITION

LARGE
CORPORATION
R&D
& TECHNOLOGICAL   STRUCTURE
DEVELOPMENT

SMALL
ENTREPRE-
NEURIAL
FIRM
INNOVATION

INTERNAL

DOMAIN

EXTERNAL

Three strategic parameters for technology-based companies.

project which allowed Japan to establish technological leadership in a portion of the semiconductor industry and was sponsored by the Japanese Ministry of International Trade and Industry (MITI) was an example of a national strategic move based on technology. In order to succeed, companies using technology must now think hard about their technological capabilities in the context of a very wide range of competitors and potential allies.

The figure above shows three parameters that must be considered by organizations whose products depend on evolving technology. Should the appropriate technology be internally developed in the laboratory of the organization or a subsidiary, or should it be contracted, purchased, licensed, or acquired through a joint venture or acquisition? Should the technology be used as a competitive weapon, or should competition be on the basis of other functions, such as advertising, appearance, price, with the technology developed cooperatively? Finally, should the technology be developed in the context of a large corporate research and development effort, or by a small, more independent entrepreneurial group?

Because technology is much more dynamic than it once was, engineers are less likely to live out their technical careers in a sinecure of expertise. Technologies such as computers and electronics have permeated industries to such an extent that engineers remain ignorant of them only at their own peril. At the same time, small companies and big ones, U.S. companies and foreign ones, have been learning from one another so that business acumen is increasing. It is now difficult to find a start-up company so ignorant of business that the engineer is given complete freedom, or a large company that is not experimenting with small entrepreneurial group structures. Firms are becoming more sophisticated in integrating technology into their business strategy and in making decisions as to how to handle it. Technology is somewhat like food. Individuals used to grow their own. But now technology appears through a bewildering system of sources, brokers, and distributors.

In summary, engineering is inseparable from money, people, and business strategy. Money is required to support technological efforts, whether it comes directly from the sale of products or indirectly from the government, banks, venture capitalists, or the public. People are central to technology, and the management of people from varied and sophisticated disciplines poses special problems. Finally, technology has become a part of business strategy, and the management of technological capability, through interindustry and intraindustry cooperation and communication and international relationships, has become essential to success in technological ventures. Engineers who do not understand this change in the game and learn to play it will have less control over their own destiny.

10

Regulation

*The Painful
Inevitability*

There was a day when technology ran relatively free—when feasibility, economics, and social pressure were the prime constraints. They are still significant. We have discussed feasibility and economics; in the case of social pressure, throughout the history of technology sufficient outrage has always gotten its way. Gun owners in Palo Alto, California, keep quiet about it, even though owning a gun is legal. However, we now limit technology in other ways. In the days before the Industrial Revolution societies felt little need to constrain technology through governmental process and decisions of law. The founders of the United States probably had no intention of setting the government up as a watchdog on technology. Yet in less than a century, the U.S. government saw fit to interfere with business to control technology. The problem that led to regulatory action was exploding steamship boilers. The chain of events is instructive.[1]

In the early part of the nineteenth century, Fulton's first steamboats traveled 3 or 4 miles per hour and were powered by Watt's steam engines, which had been proven over years of use. As steamboats were improved, they became an extraordinarily successful business. Fulton's first New Orleans boat resulted in a net profit in its first year of $20,000 on an initial investment of $40,000. When profits of this magnitude are possible, competition develops quickly, and one of the factors that gave one an advantage in this competition was speed.

Higher-pressure engines provided greater speed, since the higher the pressure, the smaller the pistons and other engine parts could be for the same amount of power; the resulting lower weight would yield better ship performance. However, owners and operators were tempted to operate even these engines at pressures above design levels and to lower costs by compromising on maintenance and safety margins. The result was a rash of boiler explosions. In the period 1841–1848 approximately 70 marine explosions killed 625 people. In December 1848 the commission of patents estimated that in the period 1816–1848, 233 steamboat explosions had killed 2,563 people. In 1850, 277 people were killed in boiler explosions. In 1851 the number of fatalities rose to 407. Although this may seem small compared with the 50,000 people who presently die each year in automobile accidents in the United States, it was too high a number for the U.S. public of the time. In 1852

Congress created the Joint Regulatory Agency of the Federal Government. This agency, initially operating through inspectors, was successful in lowering fatalities 35 percent in the following eight years. Eventually, uniform boiler codes were adopted, and boiler explosions on boats became a thing of the past.

As one would expect, tremendous controversy was associated with the creation of this agency, and bitter opposition was launched from those in the steamboat business. Then, as now, technology was closely associated with business, and government regulation of business was economically and philosophically extremely distasteful to those who believed that all things good come through the free market. In this case, all competitors were affected by the legislation, so that the competitive climate was at least uniformly changed. Today, we face more difficult problems because some competitors, particularly foreign companies, are not subject to regulation (such as pollution restrictions on factories). Then as now, there was tremendous objection to this new watchdog role of government.

Since 1852, as the need for regulation of technology has rapidly increased, so have the problems associated with it. Governmental regulatory agencies with specific responsibilities at the federal level now include the Federal Aeronautics Administration (FAA), the Environmental Protection Agency (EPA), the Food and Drug Administration (FDA), the Federal Communications Commission (FCC), the Office of Safety and Health Administration (OSHA), the Interstate Commerce Commission (ICC), the Nuclear Regulatory Commission (NRC), the Consumer Product Safety Commission, and the National Highway Traffic Safety Administration. At the state and local level many other agencies directly or indirectly regulate technology (including air pollution control boards and zoning boards). In general these government organizations regulate technology in areas where the governments have detected problems of safety or quality in the products purchased by the public.

Engineers themselves, through their technical societies, have often formed agencies to provide standardization and quality control. For example, the American Society of Testing Materials (ASTM) plays a role in ensuring the consistent quality of components and materials used in the products of technology. Another example is the American National Standards Institute (ANSI),

A model of the proposed Scheme Z interchange in Cambridge, Massachusetts, across the Charles River from Boston. This spaghetti bowl of highways and ramps is the highly controversial part of Boston's $5 billion Central Artery/Third Harbor Tunnel Project, which will sink underground the interstate highway that goes through downtown Boston. Citizen and environmental groups, as well as the City of Cambridge, filed lawsuits to halt construction of what would have been one of the world's largest—and some say ugliest—traffic interchanges.

which among other things sets safety standards. Regulation is also accomplished through standards set by purchasing groups to ensure the quality of their purchases. You have perhaps encountered the military specification system (Milspecs), which define the characteristics of a large portion of the supplies and equipment bought by the U.S. military.

One of the many reasons for this continually increasing regulation of technology is the destructive potential of modern technology. A bewildering array of products and by-products often end up not only polluting our environment but permanently altering it. The scale of technology has also increased manifold, and that is another reason to regulate technology. We have larger factories and larger products. The death toll in the crash of a single 747 can be comparable to the yearly death toll from boiler explosions that once caused the public to demand government intervention. To take another example, modern logging equipment, sawmills, and paper product plants allow us to cut down forests at an alarming rate. A third source of pressure in favor of regulation is population density, which exposes large numbers of people to any potential danger, while our unprecedented communication system daily relays the details of every major accident, disaster, and danger.

Finally, we regulate technology because our expectations of technology are increasing. We want ever-better performance from products with a growing margin of safety. We want cheaper air travel with fewer accidents, lower-cost building materials with more protection of factory workers, and more complex manned space missions without cost overruns and failures. We want sophisticated technology, but we do not want the associated unknowns. We want increasing marvels with decreasing risk.

The perception of risk is a complex and critical factor in our social acceptance of technology and therefore in our regulation of engineering. The willingness to take a risk is affected, first of all, by whether one perceives the risk as voluntary or imposed. Chauncey Starr, an engineering educator and one of the founders of the Electric Power Research Institute, hypothesized that if a risk-prone situation, such as the building of a nuclear plant in our immediate neighborhood, is imposed on us, we expect that risk to be similar to the risk we associate with natural disasters over which we have no control, such as earthquakes, typhoons, and tornados. These kill on the average of one person per million per year.[2] On the other hand, if we voluntarily undertake a risk-prone activity, such as hang gliding or jay walking in New York City, we seem willing to tolerate a risk level on the order of that associated with dying from natural disease, even though this "voluntary" risk is at least 1,000 times higher per hour of exposure than is the "involuntary" risk. If you drive a car, your risk of being killed in any given year is approximately one in 4,000, but you probably do not worry about that too much.

Applying these attitudes about perception of risk, one could therefore forecast different public reactions to, say, the use of a powerful pesticide, based on whether the use was seen as voluntary or imposed. This certainly has been apparent in California, during statewide campaigns to control infestations of the Mediterranean fruitfly, in which the pesticide Malathion has been dispensed from helicopters over heavily populated areas. Even though the concentration of the pesticide was probably weaker than the concentration many people voluntarily used in their backyard gardens, there was tremendous public outrage. The public perceived these campaigns as something imposed on them without due process. Much of technology seems to be viewed in this light. Although we consumers play a large role in shaping technology, we seem to consider damage that results from it to be "involuntary." We are far less angry if we get into an automobile accident because of a mistake we make than if the accident is caused by a failure of our automobile.

Another factor that markedly influences our willingness to take risk and accept failure is potential benefit. As benefit increases, people are willing to move to higher and higher risk levels. According to Starr, as benefits become appreciable, people are willing to accept more risk than that corresponding to natural disasters. But the benefit has to be great indeed for us to accept a risk equal to the mortality rate from disease. Since the benefits of technology are large (at this point we probably could not survive without them), one might think that we would accept considerable associated risk. However, we tend not to be aware of the benefits of technology on a day-to-day basis. We take our technology for granted. Perhaps if we could live in a world without technology every other week, we might have a better perspective on the cost–benefit ratio.

A third factor that influences the public's attitude toward risk is the perceived magnitude of failures. A number of small failures are tolerated more easily than a large one of the same cumulative size. In the United States approximately 100,000 single-fatality accidents occur for every one that kills 100 or more people. But the large ones have a correspondingly greater effect on noninvolved individuals.

Our attitude toward failure is affected by our awareness of failure, which is in turn affected by the intensity of media coverage. In the late 1970s, when Skylab reentered the atmosphere, the danger was hyped by the media into the status of an international crisis, even though the risk of personal injury was exceedingly small. The DC-10 accident mentioned in Chapter 7 almost caused an innocent airplane to be executed by the media without a fair trial. The newsworthiness of airplane crashes continues to keep the public apprehensive about air travel, even though air travel continues to be much safer than automobile travel for an equivalent number of miles. We are very sensitive to the risks associated with technology because of these characteristics of risk perception. We want to make sure that technology does not hurt us, but we are very ambivalent about failure even *without* risk to life, limb, and property. As a culture, we are success-oriented, and we value those who take risks and win. We grow up listening to adages such as "Nothing ventured, nothing gained," and as adults, and particularly as investors, we hear "The greater the risk, the higher the return." Risk-takers who succeed reap social and economic rewards. We certainly admire the Silicon Valley entrepreneurs and the great historical figures of technology. But what happens to the business risk-takers who fail instead of succeeding? At the very least, their bankers turn several degrees cooler. How about national failures? Our country is young enough and has been fortunate enough that we have suffered few setbacks. Those we have suffered (the Vietnam War, the present trade deficit and inability to balance our budget, and such technological failures as Three Mile Island, the *Challenger* explosion, and the defective mirrors on the Hubble Space Telescope) bother us a great deal. Failure contradicts our basic success orientation, and calls our national identity into question. We do not even like to think about possible future failure, and when failures happen we lurch into a period of overreaction.

At a recent meeting of executives from various companies, I was sitting at dinner with a number of people from U.S. companies and one gentleman who was a vice president of a Korean company. Because of the age of some of the people at the table, the talk turned to the Korean war and Korea's recovery since the war. The tone of the comments of the U.S. businessmen was a bit patronizing, until it turned out that the Korean company had a business volume far in excess of that of any of their companies. At that point, more interest was shown in the opinions of the Korean executive. During the ensuing conversation, he was asked what Korean businessmen thought of American industry. His reply was that it had "great potential." At that point, needless to say, a silence fell. The Korean was viewing our economic success from his own vantage point. To him, U.S. industry simply had not been tested by adversity; no modern wars had been fought on our soil, we had not withstood an occupation by another country, we had weathered no coups nor could we speak of our history in thousands of years. He saw no reason why such adversities would not happen to us just as they had to other nations, and these would be the real test of our resiliency and creativity.

We in the United States have trouble with this point of view. We would rather think about our successes rather than about past or future failures. An occupation or a coup are unthinkable! We assume that all ventures should lead to success, and if one does not, lawsuits and regulation often follow.

Our ambivalence toward (and misunderstanding of) risk and failure has a profound effect on technology, which necessarily includes both. Our ability to be innovative depends on our ability to experiment, and with experimentation comes failure. In the cases of large projects, these failures sometimes are worrisomely intolerable to our society. A good example of this could be seen during the "energy crisis" of the 1970s, when the nation was wisely (and temporarily) facing up to the fact that petroleum is a finite resource. In the long run, we must broaden our energy base. Whether the 1970s was the time to start or not is still a subject of debate, but at the time it seemed that it was, and the resulting public reaction was typical of how our society presently views changes in technology. We found that every alternative energy source we considered involved risk. By now we are all aware of the risks of nuclear power, particularly the problem of disposing of radioactive by-products. However, we found that a shift to coal would result in a higher death rate and lost-time accident toll than that associated with conventional oil or nuclear energy production, because of mining and processing hazards. Increased use of coal would also mean that we would have to learn to heal the major scars on the land that result from economical forms of mining. Solar energy, including wind power and the conversion of biomass, would require large amounts of raw materials and often extensive land areas. Such approaches as the production of alcohol from field crops and the direct conversion of trees would require very large-scale single-crop farming, with accompanying worries about erosion, land exhaustion, fertilizers, and pesticides. Each way we turned there seemed to be risks.

Major changes in the energy pattern will have enormous economical and ecological ramifications when they occur. There will be no free energy lunches. Attempts to improve efficiency result in greater sophistication and complexity, with accompanying potential for failure. No matter how careful the players, mistakes will be made and failures will occur.

The public reaction to this new knowledge was bafflement. Unfortunately, we saw the risk in the move to alternative energy sources from the most cynical viewpoint. We had difficulty seeing such a move as voluntary; rather than seeing diversification of the energy base as something that should be done because of resource limitations and changes in the world economy, we preferred to believe that such a move was being imposed upon us by unfriendly foreign interests, the petroleum industry, Democrats (or Republicans), unenlightened or selfish interest groups, large cars, stupid politicians, or whatever. Further, the rewards were subtle. Efforts at conservation, for example, would not add to our affluence; at best they would allow us to slow our rate of loss. This is hardly an inspiring reward for high rollers. Since the early 1970s, we have bounced up against the energy problem several times—it is not going to go away—but we remain baffled, incapable of acting.

While we are very aware of failures and potential catastrophes, we are amazingly ignorant of what we have and how it works. We have become strongly sensitized to gas leaks, oil spills, rolling brown-outs, and losses in nuclear material—problems felt throughout the United States. Yet how much do most of us know about our overall energy situation? Do we know what percentage of our energy comes from which sources, and how much is used

A forest of Fraser firs on Mt. Mitchell in the Appalachian Mountains. The trees were purportedly destroyed by acid rain. The standing dead can be seen shrouded in acidified clouds from industrial and automotive pollutants.

by various groups? Do we know how many central power plants there are in the United States? How many public utilities? Do we know how much investment is needed in different aspects of energy production? Or how much time is required to make major modifications in a major system? Since we do not want failures, as the energy "crisis" recurs this bias in our information base causes us to be extremely conservative. So on the one hand we view ourselves as a leading technological power capable of solving problems, while on the other hand we are extremely conservative in our willingness to apply technology to a solution of the energy crisis. Such dichotomies lead to great tensions and complexity in the regulatory process.

Public attitudes toward technology are based on many things other than logic—on the desire for something for nothing, on wishes and dreams, on factors distinctly human rather than distinctly technical. Regulation reflects these attitudes. It therefore leads to a large amount of controversy and argument among the various represented interests. Regulation of technology represents a political statement as well as a scientific statement. It represents hopes and dreams as well as practicality. Keeping that in mind, let us look at various ways in which technology is regulated.

A major regulator of technology is simple consumer demand. It was not "in" to own an Edsel. Home dental extraction kits would not sell well. But technology is also regulated through social "pressure." I own a motorcycle with a noisy (although probably legal) muffler. I am careful to handle the throttle gently when driving through residential neighborhoods, especially the one in which I live. The process by which the marketplace and social values regulate technology, however, is not as simple as it used to be because of the existence of modern media and the biasing of social values through mechanisms such as advertising. A good example is the continued advertising of cigarettes, which have been proven to be extremely harmful to health. It is probably safe to say that few people have particularly enjoyed smoking their first few cigarettes. Even long-time users seem to wish that they did not smoke. But the cigarette industry has so successfully identified their product with various attractive images that young people continue to take up smoking cigarettes, and eventually become dependent. The market for cigarettes became distorted to the point where public resort to regulation of cigarette advertising, package warnings, and nonsmoking zones was inevitable.

Another example of how advertising increases natural demand is our present love affair with personal computers. It is "in" to own computers, to be computer literate. In fact, one should own a fairly up-scale computer. I would not have typed the manuscript for this book on a TRS-80 or other early machine for fear that one of my friends would find out. Therefore, I typed it on a machine with 5 megabytes of internal memory, a 40-megabyte hard disk, and a word processor program most of whose features I neither need nor know. Each day I am bombarded by advertisements of new and far better machines offering me greater speed and capacity, and software offering me windows, greater format capability, dictionaries, and other such wonders which I also do not need but may buy. If this trend continues for too long, there may once again be a backlash, this time against the "computer revolution," possibly with a demand for something akin to "truth in advertising."

It is ironic that business, which puts enormous emphasis on regulation by the market rather than government, expends its resources to push demand above its natural level, which then leads to a greater need for government regulation. For example, if the automobile business had not been so successful in luring us into cars, we might not have the amount of traffic, air pollution, and safety regulation we now have. If the airline industry had not been able to get so many of us into the air by their use of pricing gimmicks, give-aways, and images of exotic vacations, we might not have had the chronic schedule slippages that caused the FAA to intervene. If it did not advertise luxury, we might not be so upset with lines, delays, and lost luggage.

Some analysts think that if the market and public pressure do not effectively control technology, industry should do so. Here we have a built-in conflict of interest. Most private industry sees its goal as the making of profit. Technology makes money, first because it helps lessen the costs of production, and second because technology is incorporated into products. Profits can be increased if some of the costs of using technology, such as adequate disposal of pollutants and the provision of safe working conditions, can be ignored. There is also the temptation to sell whatever technological marvel the short-term market will buy, regardless of the long-term costs to society. Most managers I have met from private industry are bright and well-meaning people. However, their view of public welfare is tightly coupled with economic growth, and they see their own top priority as the health of their business. "What's good for General Motors is good for the country." The same can be said for managers of government enterprises. We only have to look at the newspapers, engrossed with such issues as contamination in government fissionable materials plants and pollution within military reservations, to see similar indications of their desire to accomplish short-term goals while neglecting long-term ones.

How about the individual engineer? Should he or she be the watchdog?

Once again, the answer appears to be partially, at best. There is presently much ado about engineering ethics and the responsibility of the individual engineer. Many engineering societies have adopted statements of ethical responsibility. But the situation is complicated. For instance, the codes ask the engineer to be responsible for the public interest. However, most engineers are employees of organizations to which they necessarily must have some loyalty if it is to function. Organizations also hold power over engineers, since they can influence their paychecks and their future employability. The engineer can easily become caught in the middle, between public interest and self-interest. Second, many engineers do not see themselves as watchdogs. The life of a typical engineer is filled not only with attempting to do a good job at work but with raising kids, paying for a house, attempting to take a vacation, and the other things that people do. To add a responsibility for the behavior of technology to this would be considered by many to be asking a bit much.

Finally, even if one were willing to take on the job, making decisions about ethical and moral issues is not easy. What is the "right" thing to do if the company you work for accepts a contract to develop a new nuclear weapon, or is dumping chemical wastes into the ocean? And even if you blow the whistle on your company's misconduct, can you be sure the legal system will back you up? What role does the legal system play in determining rightness and justice in cases of dispute over proper conduct? If a talented attorney spellbinds a jury into a verdict that would let a small company off the hook because to clean up a twenty-year-old toxic waste dump would bankrupt a family business, is that verdict by definition "just"?

So far whistle blowing has not been a positive experience for those blowing the whistle. Three Bay Area Rapid Transit engineers who went outside of channels to report safety problems were fired for insubordination, incompetence, lying to their superiors, causing staff disruptions, and failing to follow understood organization procedures. Two years later they sued BART and settled for damages out of court. Later they received the Award for Outstanding Service in the Public Interest from the Institute of Electronic and Electrical Engineers, but the short-term cost was high.[3] Three engineers from General Electric who resigned because they could not find a company forum for expressing their concern over the dangers from nuclear power generation similarly suffered inconveniences.[4] There are presently ongoing attempts to protect the rights of individuals who follow their own ethical motivations, but such actions alone will probably also not be strong enough to solve the problem.

All of this is not to say that individual engineers should not take on more ethical responsibility. They are often in a position to detect shoddy practice and dangerous directions before others can. Regular and short courses on ethics for engineers are becoming more common, and the public is becoming more supportive of engineers who take a strong stand against unethical practice. However, individual engineers can provide only a part of the watchdog function.

We reached the point with the exploding steamship boilers where certain social values pertaining to technology had to be made into law. That legal process has been occurring ever since. Some laws give us little trouble, such as the illegality of driving cars without mufflers, buying crack, or selling DDT in grocery stores. Other laws, such as those having to do with motorcycle helmets, gun control, and seatbelts seem to give some of us a great deal of trouble. Some laws have tremendous strength. The provision of the Geneva convention which required full metal jackets for bullets used in war is still honored, even though megaton warheads, nerve gas, fully automatic weapons, and aerial delivery of all sorts of horrors have been used since. Other laws, such as those prohibiting use of marijuana or driving above 55 mph, seem to have no strength at all.

As time passes more and more laws are passed to regulate technology, and more and more government and private legal action is taken to interpret those laws and establish recompense and punishment. Unfortunately, the regulatory process involving law and the courts is a complex and increasingly time-consuming and expensive part of life. It is generally an adversarial one and involves engineers in many roles. A good example is the case of nuclear power in California, a technology that was halted after many years of battle.[5] A number of specific sites were involved in the fight; plants were planned for Bodega Head on the coast north of San Francisco, Malibu on the coast north of Los Angeles, and Diablo Canyon on the coast by Santa Maria. The

Vallecitos atomic plant was already in existence. Each of these sites became the subject of vicious contests among the public utilities, private companies, the Atomic Energy Commission (AEC), the Sierra Club, a number of other environmental and property-owner groups, the Public Utility Commission, the United States Geological Survey (USGS), and a large number of independent consultants. The first two plants were canceled and the Diablo Canyon plant has gone through an agonizing time. The Vallecitos plant, which manufactured medical isotopes, was closed for six years, during which it lost its business to sources in other countries. The basis of the argument in each case had to do with the safety of a plant in case of an earthquake. The viciousness had to do with the very different perspectives of the various parties involved.

For example, the expertise of both engineers and geologists was needed, but their perspectives were quite different. Engineers are used to designing for dangerous environments, but most of these environments change rapidly enough that data can be gathered to describe them rather closely and make predictions useful in design. Geologists, on the other hand, are concerned with very large periods of time and work with data that are subject to great differences of interpretation. At the Bodega Head site the USGS geologists concluded that a small fault had seen some movement, perhaps one to three feet, some time in the last 40,000 years. The engineers were able to deal with small movements, but they needed to know the expected movement within the next 50 years (the lifetime of the reactor). To a geologist, 50 years is not a meaningful period of time. The process of forcing a historical science such as geology to converge with a field such as engineering proved extremely difficult.

Environmental and property-owner groups were opposed to the planned reactor on a variety of grounds, ranging from disruption of the sea coast to complete opposition to atomically generated energy. The utilities wanted the electrical production of the plants, and the private companies wanted the construction business. The AEC had a decision-making role and was still affected to some extent by its post World War II mission of "Atoms for Peace." Governmental agencies are also influenced by governmental priorities, which in turn respond to the electorate and various interest groups. The USGS was a group of highly competent professional geologists on the government payroll, who of course had their own individual attitudes toward atomic energy. The consultants were retained by one of the adversarial parties, often not only because of their professional competence but also because they had similar attitudes toward nuclear power.

In the adversarial process, technical experts with equivalent credentials often, if not usually, appear on both sides of the case. The complexity of issues involving technology often forces such experts to rely on their own values and opinions. As a beginning professor, I consulted as a technical expert in a few legal cases, two of which came to trial. Obviously there was no simple answer to these arguments, or else the trials would not have occurred. We experts were therefore offering opinion, and the debate swung around to our credentials and credibility. I found myself in the interesting situation of sitting in public while a very talented and articulate attorney tried to prove that I was not very competent. I did not need that, and have tried to avoid courtrooms since. Many engineers enjoy working as an expert witness and develop an appreciation for and expertise in the proceedings of law. Others find it extremely frustrating. Many trials were conducted over the use of nuclear energy in California.

Not only was there litigation, but a number of government hearing boards, such as the Nuclear Regulatory Commission (NRC), the Atomic Safety and Licensing Board (ASLB), and the Advisory Committee on Reactor Safeguards (ACRS), would review proposals and objections and gather evidence for their decision. There were also formal appeal procedures from their decisions. At the time of the California site battles, applicants for permission to build a reactor were planning on a minimum of three years from the time they applied until they received a decision. The California litigation and widespread debate extended that even more.

Such long, involved procedures are not only expensive and divisive, but are also hard on U.S. business. Companies in other countries, such as France and Japan, build reactors with much less public input. Topics such as reactor siting and safety in France are debated behind closed doors and do not include public hearings. The regulatory process in the United States has prevented nuclear reactors from being competitive either with conventional power plants or with reactors built in other countries. To those who do not want to see nuclear plants in the United States, this outcome is reasonable. To those who point out that the United States has given away a lead in technology and a commercial business to other countries and who feel that nuclear power is safe, the outcome is not reasonable.

While at JPL and later at Stanford I participated in and led studies on air pollution monitoring, and I was a member of the California Governor's Task Force on Toxic Waste Disposal. I noticed that the nontechnical people on committees having to deal with such problems, like most people, would like perfect technology, which costs next to nothing and will not result in future unknowns. Generally, technology already exists that is better than the technology in use, but it is expensive. New technology can be developed but is even more expensive and will result in even more unknown side effects. Engineers involved in these debates tend to become frustrated by the expectations of the nontechnical people and unsympathetic to their unwillingness to pay for new developments. They also tend to be uncomfortable with the squishiness and squeamishness of the debate.

In the case of the Governor's Task Force, the engineers wanted an answer to the "How clean is clean?" question. The control of toxic wastes requires a many-pronged attack. First of all, existing toxic waste must be cleaned up. Second, the amount of waste that is generated must be reduced. Finally, better methods of disposal must be found. However, unlike the air-pollution situation, there are no standards for acceptable contamination in the ground. Until a target is set, engineers have difficulty making technological proposals. But for various political reasons, no targets were set.

The engineers also wanted more state money allocated to research and development on toxic waste disposal. However, the state administration, being on the conservative side at the time, felt that technology should be developed through the private sector. Toxic waste disposal is a very complex field, and therefore private companies are reluctant to make large investments in advancing the technology, in light of the unsure payoff. The differences in perspective between the technical and nontechnical people participating in the study often led to disappointments because the solutions the nontechnical people wanted were not forthcoming, nor was an appreciation of the technical difficulties forthcoming, as the technical people desired. In fact, the major stumbling blocks to toxic waste disposal in California are legal and political, not technical, a situation that annoyed both sides. The technical people were a bit miffed to see that they did not hold the key to the puzzle, and the nontechnical people were hoping that the legal and political problems would not have to be solved.

Part of the reason that these problems have still not been solved is the high cost of the political and legal wrangling. For some years now, the Rand Corporation has been conducting a project entitled the Institute for Civil Justice in which they have been studying the judicial process as it applies to damage suits of various sorts. They have come to some interesting conclusions.[6] For instance, in the average nonautomobile tort lawsuit in 1985, plaintiff litigation expenses were 31 percent of the total compensation, and defense expenses were 28 percent. The persons wronged received only 41 percent. Although these cases include some having to do with areas such as assault, libel, slander, and land disagreements, they cover suits for personal injury or property damages and injuries resulting from negligence or breach of warranty. The study does not break out the cases directly involving technology. However, expenses are large, since the study estimates that the total annual expenditures for tort lawsuits in the United States in 1985 was somewhere between $29 and $36 billion. An even more startling set of numbers can be seen in the case of the presently ongoing asbestos litigation. Over the past ten years, an estimated $1 billion has been spent on compensation and litigation expenses. Since estimates of the number of deaths due to asbestos over the next 30 years range from 74,000 to 265,000, it is expected that the final level will be many billions of dollars. So far, plaintiffs have on the average netted 37 percent of the money spent by all parties on such cases. Defense expenses have averaged 37 percent and plaintiff expenses 26 percent. The legal and court costs are almost twice the compensation to those wronged. Since both the number of tort cases and the average awards are rising sharply, the expense of using the courts to determine deviance from regulatory standards and penalty is obviously something that our society must confront.

The Rand study also discusses the difficulty of submitting sophisticated technological arguments to a jury. The following are some comments of jurors after hearing expert witnesses:

"Not that he didn't seem smart enough to lie, but he was more . . . he was believable."

"To me there was nothing fake about him."

"He was not rehearsed."

"He was an honest man, you might say."

"He's a good old boy."

"You knew he had been briefed and he had gone over this, but it wasn't like some of the others. Some of the others, you wondered how many . . . they'd probably get up at night saying what they said . . . You've got all of them in there in front of the mirror and combing their hair saying what they were going to say."

These comments have little to do with expertise.

The regulation of technology is a complex issue, but we can and will do much better at it. Nevertheless, it will take a better dialogue and more education among the various parties involved. In a simpler time, we handled the issues through the adversarial approaches of the law, and we are still trying to do so. However, today this costs a great deal, strains the judicial system, and yields very uneven results. Even the traditional adversarial relationship of business and government does not seem to work any more. Even though business executives tend to be against government regulation, they seem to prefer it to individually initiated programs (which put the initiator at the disadvantage of competition). Many U.S. executives who were loudly in favor of the free market have done a turn-around now that Japan is such a fierce competitor. It has become increasingly clear that technology has reached a stage of maturity where social control of it is essential to both our economic and our physical health.

11

Thinking Technology

The Challenge for the Future

Speculating about the future is not my business, and so this chapter will be brief. I sympathize with people who attempt to make a living by predicting future technological trends, for two reasons: they are often wrong, and they are up against strongly held individual values. I am privy to many heated discussions of technology, and although they may be dressed in data and apparent logic, they are usually quite similar to arguments about religion or politics. I recently overheard two of my well-informed faculty friends discuss the pros and cons of nuclear energy, each using the same data on nuclear waste to justify his argument. Not surprisingly, my politically liberal friends, many of whom labor in the humanities, tend to be more suspicious about technology than my politically conservative friends, most of whom are engineers and business people. I am a middle-of-the roader when it comes to technology. I don't believe we are going to destroy ourselves through technology, nor do I think that technological breakthroughs will bring us utopia. The extremists among my liberal friends therefore consider me dangerously pro-technology, and my archest conservative friends view me as pathetically pessimistic. Neither group would give my predictions much weight. Nevertheless, it seems to me that some trends can be discerned from analysis of our present achievements and difficulties.

For instance, it is fairly safe to predict that many of the hardware and software wonders predicted in books about the future will occur. The present rapid development of information and communication systems will continue for a while, with improved satellites, fiber-optics interconnections, universal cellular networks, and increasingly sophisticated and miniaturized equipment. Biotechnology, or genetic engineering, will continue to bring us new pharmaceuticals and will infiltrate many nonmedical aspects of life. Unfortunately, most of the public's attention to date has focused on the advantages and potential dangers of using genetic engineering to treat human diseases and to alter plants and animals in the food chain. Its potential in the basic process industries is often overlooked. Fermentation is one of the oldest and nicest chemical processes. Naturally occurring bacteria do it quietly without large amounts of electricity or barrels of fuel oil. Bacteria that have been genetically altered are capable of a much wider variety of processes. Such microbes already eat oil spills and concentrate ore and can be engineered to do much more. Perhaps, as one of my students said, genetic engineering will be "real" when it can make better beer.

We will have such things as space stations, perhaps a scientific colony on the moon, and probably weapons in space. There is no indication that disputes and wars will cease, or that the military and public fascination with "star-wars" technology will lessen; so, economics permitting, the battlefield will probably eventually move to space. The competition and the fireworks would be maintained, and the number of human lives at risk could, and necessarily would, be limited, or so the theory goes.

But as we become more sophisticated in our technology, we will, I believe, also become more sophisticated in the manner in which we use it. For instance, in the immediate future I expect much more thought to go into questions of appropriate use of our new-found capability in communications and information processing. How do we balance accessibility and privacy? How do we separate news from entertainment, or do we? Who needs to know what? Technology is advancing by leaps and bounds, but human brains are not. Many of us are already saturated with information to the point of confusion. How do we sort through it all?

I remember feeling strongly about this necessity during the 1990 election. The ballot in California was very complicated, with many special-interest measures, each of course written to mask its true goals from the voter. Since these issues were important to industry (restraints on pesticides and logging, taxes on liquor), a great amount of money was spent on radio and TV to rally support for or opposition to these measures. The advertising spots were so well done and delivered such an emotional impact that it was almost impossible to think clearly about the measures. I would watch a woman with a young child bald from chemotherapy talking about the unknown risks of pesticides and I would vow to love the pests in my garden. I would then watch a poverty-level family facing the increased costs of an improved environment and feel like chopping down a redwood. To make matters worse, as I headed for my polling place, the computers were once again predicting results, even though they were supposed to wait a discreet interval. When the details of our democracy were put in place, modern electronics did not exist and public opinion was not formed by television. Some thinking is surely needed about what technology is doing to the democratic process.

We will become more profound thinkers about the uses and directions of technology not only because of its increasingly powerful capability but also because life, or at least our awareness of it, is becoming more complex. For instance, until the middle of this century, engineers were able to ignore all but the economic aspects of limited resources. This is no longer the case. The world's population continues to increase at a rapid rate, subjecting not just our mineral resources and rainforests but also our atmosphere, oceans, and cities to unprecedented stresses. Such considerations become major engineering constraints. Questions of energy efficiency, recycleability, and appropriate use will become ever more important as the quality of life in the crowded areas of the earth degrades and as shortages and negative health effects become noticeable.

For example, oil and coal used as fuel have allowed us to work wonders, but they are too valuable as complex hydrocarbons that can be converted into all sorts of other forms (such as plastics) to be so rapidly burned in automobiles, power plants, and furnaces. We are also very aware of the negative effects of this burning upon the environment. In the future we will see increased use of photovoltaic generation, or direct conversion of sunlight to electricity, as the cost of the conversion equipment falls. The development of higher temperature superconductivity will also become a significant factor in increasing the efficiency of energy usage.

In the short term, I predict tremendous pressure in the United States for a return to the nuclear energy business. Fission power is some time in the future, but it will eventually become available, while fusion is already much more heavily used in other countries. France, for example, generates over 70 percent of its electricity by nuclear means. In the United States, no new nuclear plants have been ordered since 1976. As time passes, problems having to do with plant safety and waste disposal are being solved while environmental problems from petrochemically fueled power plants seem to increase. We will necessarily become more sophisticated in our thinking about nuclear energy in order to deal with this pressure.

The explosive demolition of Pruitt-Igoe in 1972—a low-income high-rise housing project in downtown St. Louis that turned into a vertical urban ghetto. Ironically, the tenants found the buildings well designed in some respects; had they been built in a different socioeconomic environment, these slabs probably would have provided acceptable housing.

The old laissez-faire approach to technology has run its course. We can no longer afford to produce anything we happen to dream up, and some things we can afford have unacceptable side effects. Society will have to make some hard choices about the directions of technology, and at present adequate mechanisms for making these choices do not exist.

Many of our social and environmental needs can be met only by applying high-tech thinking to low-tech problems. Unfortunately, glamor areas in technology tend to attract a disproportionate share of talent and resources. Yet we need improved transportation systems that move people rapidly, comfortably, and efficiently, better housing at a moderate price, adequate fresh water for population centers, and more affordable medical care and education. We need higher quality mass entertainment and less garbage and better means of disposing of it—products that can either be recycled or repaired rather than consigned to the land-fill. We need cleverer ways of maintaining our aging roads, bridges, and pipes, and of having our forests and timber products too.

How are these kinds of changes likely to affect the professional lives of engineers? Increasingly complex technology and social problems will result in more constraints and requirements and require stronger interaction between the community of engineers and the larger society. Engineers must learn to produce better products, which not only function well but satisfy emotional needs, are consistent with cultural values, fulfill our intellectual desires for quality and sophistication, and have minimal adverse effects on the environment. Many present products accomplish one or two of these but fail on others. My vintage Jaguar pleases me emotionally, but it breaks down and does not like to pass its emissions check. I have an American-made pickup truck that is highly functional and passes its emissions check, but its paint is peeling off and it's not much fun to drive. My new stereo receiver makes wonderful sound, but I can't seem to memorize its controls well enough to operate it when the light levels are too low to read the tiny printing on the panel. Commercial airliners are beautiful and fast, but I don't fit in my seat, and I find that I can drive to Los Angeles almost as fast as I can fly if I count commuting to and from airports and the required airport rituals of parking, waiting, and standing in various lines. My chain saw, power lawnmower, and weedcutter are convenient but make a horrible noise and require a lifetime of engineering experience to keep them running well. Telephones seem to be necessary, but calls from the East Coast wake you in the morning, and local entrepreneurs call up during dinner to sell you stock or get you to contribute to the police. We engineers must do a better job of invention and design in the future than we have done in the past. Technology is supposed to improve our lives, not use up our time and money and leave us unhappy.

Mathematics will continue to bring engineers new tools and understanding, and computers will continue to increase their ability to analyze, simulate, and optimize as well as to combine graphic with analytical work and to access information. Science will point in ever new directions that will lead to improved products based on deeper understanding of physical behavior. Chemistry will provide new insights into materials, processes, and medicine; physics will continue to contribute to the electronics revolution, materials development, and energy generation. Meanwhile, biology will find applications in areas as diverse as agriculture and computer memory, and geology will remain crucial to our supply of natural resources, ranging from oil to water, and to our ability to build structures that can withstand the forces of nature.

Engineers of the future will still make mistakes and resort to experiments and tests to handle complexity, unknowns, and limitations in theory and understanding. However, increasingly sophisticated computer simulation and more powerful approaches to data handling should decrease the necessity of full-scale and prototype testing. Manufacturing and assembly are even now undergoing great creativity and change in the United States, after years of neglect. Flexible assembly is displacing traditional mass production, computer control is displacing human control, and reliability is going up while cost is going down. Industry does seem to be following agriculture in mechanization and automation, with unknown long-term effects upon society.

As for business and money, the last few years have seen a global move toward capitalism and concern with economic competition, which will couple technology even more tightly to business. The less-developed countries now have access to technological knowledge and educated engineers. Their continual striving to increase their standard of living and their abilty to compete internationally, coupled with the desire of the developed countries to take advantage of cheaper labor and new markets, will increase the amount of engineering activity that cuts across national boundaries. But it will also further complicate the game, owing to the ambivalence of countries such as the United States. Our government wants to help less-developed countries become economically stronger not only for idealistic reasons but also so that they will be political friends and importers of our goods. However, we become annoyed if they become so strong that they beat us in the marketplace or defy us politically. In the years ahead we will see more and more debate about technology policy—how should the U.S. government act in order to best maintain this precarious balance, if indeed we conclude that such a balance is critical. The United States has very little technology policy to date. Undoubtedly we will eventually have more, and engineers will be deeply involved in its formulation.

As a nation we must become more proficient at regulating technology, at home as well as abroad. Our regulatory bodies need to be better informed about the technology and science that fall within their domain, but they also need to be attuned to the wishes of the larger society. Our mechanisms for establishing a regulatory standard, for enforcing it, and for punishing offenders are primitive. The present legal system especially is cumbersome and expensive and not sufficiently sophisticated when it has to deal with matters of science and technology.

Finally, to be effective we as consumers and citizens are going to have to better understand our technology in the future. Our society has been delinquent in keeping up with it. One of my sons recently bought an old tractor and has been restoring it. The ignition system was giving him problems, so I lent him a copy of an old manual that had belonged to my grandfather and was the mechanic's bible of the 1920s and 1930s (*Dykes Auto Manual*). The manual described the ignition system of his tractor in great detail, but it was not particularly easy reading for him or for me. I know that my grandfather and his friends, none of whom had much formal schooling, used this book to maintain their own automotive equipment. At least as far as automobiles were concerned, I would guess that the technical sophistication of the product was lower, but the relative knowledge of the owner was higher. Since then, the sophistication of technology has increased many fold, but if anything the technical knowledge of the general public using that technology has decreased. This trend is not consistent with a safe and successful future.

I can think of several ways to increase our understanding of technology. Unfortunately, taking classes is not one of them, since there is currently a scarcity of accessible courses in schools on engineering and technology. The education system teaches such subject matter to the priests rather than the flock, and even the priests spend their time absorbing specific knowledge and techniques rather than grappling with the nature of the overall enterprise. Despite the frequently expressed concern in the United Staes about scientific and technical literacy, I predict that change will come slowly, because of tremendous vested interests in the educational system.

In the meantime, if you want to learn more about technology, you might consider working engineers themselves as a resource. Contrary to the stereotype, many, if not most, engineers love to talk, especially about their own activities, but they are not usually encouraged to do so. A language barrier must be overcome, but this is easy to do if it is acknowledged: It is only necessary to ask the meaning of terms which you do not understand. Engineers are used to this problem, because we spend our lives attempting to understand new things. Ask engineers what a typical day at work is like. What accomplishments are they proudest of ? What do they think of the public's attitude toward technology? Do they think that your kids should become engineers, and if so what kind?

It is also quite easy to become technically more specific. Ask them to explain in plain English what they do or what they are working on, and then just begin asking questions. "I am trying to design a better pyrotechnic pinpuller." "What's a pyrotechnic pinpuller?" "It's a latching device that uses an explosive squib for energy." "What's a squib?" "A little thing like a firecracker that is electrically ignited." "How does it pull the pin?" "The pin is attached to a little piston in a cylinder, and when the squib is fired, the explosion pushes on the piston." "I forget, is the piston the outside part or the inside?" "The inside." "What do you use pyrotechnic pinpullers for?" and so on.

Or tell engineers that you have decided that you want to become less ignorant about technology and you would appreciate a short explanation of whatever they think the public should most understand about their work, their project, or whatever. People like to talk about themselves and what they do. I occasionally teach classes to beginning engineering students who are trying to learn more about the profession. I often give them the assignment of finding engineers, following them around for a day, and then writing a paper on what they do. The students find it initially difficult to approach a stranger, and the engineers they approach are sometimes initially quite shy about talking about themselves. However, when they find that the students are really interested, the engineers are sometimes difficult to stop. The students also become much more comfortable with asking questions when they find that engineering is as much based on questions as on answers.

A large number of magazine articles and books discuss the products of engineering and the experiences and views of engineers and others in related fields. Some of the more entertaining and enlightening books are listed in the following section. They are written for the intelligent layperson and do not assume sophistication in math, science, or engineering. But many of them contain extensive bibliographies, including articles and papers for those readers seeking more specific details.

The organized study of engineering and technology is quite young. Some areas, such as the history of technology, are maturing rapidly, with attendant societies, publications, and acknowledged experts. However, many aspects of engineering, such as design and invention, do not fit the established academic disciplines very well and therefore have been less studied and written about. Consequently, the amount of written material about various aspects of engineering does not correlate especially well with their present importance. But the publishing world, like the world of engineering, does not stand still. In the future, I predict, we are likely to see many more books which, like this one, attempt to fill some of these gaps in our common understanding of technology and engineering.

1. A Brief History of Technology

1. J. K. Finch, *The Story of Engineering* (New York: Doubleday, 1960). A book of interesting anecdotes and accomplishments with lots of vignettes of famous engineers and projects.
2. R. Engelbach, *The Problem of the Obelisks* (London: T. Fisher Unwin, 1923). An old but interesting study of the design, fabrication, and erection of Egyptian obelisks.
3. G. R. Driver and J. C. Miles, *The Babylonian Laws* (New York: Oxford University Press, 1955).
4. Finch, *The Story of Engineering.*
5. G. F. C. Rogers, *The Nature of Engineering* (London: Macmillan Press, Ltd, 1983). A short and readable book on engineering which is quite parallel in philosophy to this one and also treats the relationship between engineering and science, design and creativity, and control of technology.
6. J. G. Landels, *Engineering in the Ancient World* (Berkeley: University of California Press, 1978). Fascinating information on such things as transportation, catapults, water systems, and energy sources in ancient times.
7. C. Singer, E. Holmyard, A. Hill, T. Williams, *A History of Technology,* vol. 2 (New York: Oxford University Press, 1957). One volume of a heroic five-volume compendium of information about technology from very early times to the end of the nineteenth century, including details on everything from the mummification of pharoahs to early submarines.
8. L. White, *Medieval Technology and Social Change* (New York: Oxford University Press, 1964). An extremely readable and highly regarded book covering several social–technological "revolutions."
9. J. Gimpal, *The Medieval Machine* (New York: Holt, Rinehart and Winston, 1976). An argument supporting a surprisingly high level of technological sophistication in the medieval period.
10. R. Mark, *Light, Wind, and Structure* (Cambridge: MIT Press, 1990). A beautifully illustrated discussion of the interaction between the aesthetic qualities and structure of Gothic cathedrals.
11. R. Marten, *Science, Technology and Society in Sixteenth Century England* (London: Howard Fertig, Inc., 1870).
12. S. Pollard, *The Genesis of Modern Management* (Cambridge: Harvard University Press, 1963).
13. D. Pletta, *The Engineering Profession* (Lanham, MD: University Press of America, 1984). A sociological study of engineers.

Other Books of Interest

A. Pacey, *Maze of Ingenuity* (Cambridge: MIT Press, 1985). A most readable treatment of several important technological developments beginning with the medieval period and continuing through the nineteenth century.

W. H. G. Armytage, *A Social History of Engineering* (London: Faber and Faber, 1961). A book beginning with technological "legacies" from the East and continuing through the development of technology in Britain.

E. T. Layton, *The Revolt of the Engineers* (Cleveland: Case Western Reserve Press, 1971). An interesting story of the early engineering professions in the United States, their attempt to do social good in the 1920s, and their subsequent decline during the depression.

C. W. Pursell, Jr., ed., *Technology in America* (Cambridge: MIT Press, 1981). A series of articles on outstanding individuals and ideas.

2. Beyond the Calculator

1. R. Perucci and J. Gerstl, *Profession without Community: Engineers in American Society* (New York: Random House, 1969). A study of the engineering profession and the people in it.
2. S. J. Kline, "Innovation is not a linear process," *Research Management*, vol. 28, no. 4 (July–August 1985): 36.
3. "Can Britain's EMI stay ahead in the U.S.?" *Business Week*, April 19, 1976, p. 122. "The 1979 Nobel Prize in physiology or medicine," *Science*, vol. 206, Nov. 30, 1979, p. 1060; "GE gobbles a rival in CT scanners," *Business Week*, May 19, 1980, p. 40. "EMI and the CT scanner," Harvard Business School Case 383–194, parts A and B (1983).

Other Books of Interest

W. Vincenti, *What Engineers Know and How They Know It* (Baltimore: Johns Hopkins University Press, 1990). Several examples of engineering accomplishments used to illustrate the problem-solving approach of engineers.

R. McGinn, *Science, Technology, and Society* (Englewood Cliffs, NJ: Prentice Hall, 1990). An overview of the field of studies known as STS.

L. Hickman, ed., *Technology as a Human Affair* (New York: McGraw-Hill, 1990). An excellent set of readings on the relationship between technology and society.

S. Florman, *The Existential Pleasures of Engineering* (New York: St. Martins

Press, 1976). A book of minor-classic status emphasizing the joy in technology and rebutting antitechnology movements.

J. D. Kemper, *The Engineer and His Profession* (New York: Holt, Rinehart and Winston, 1975). One of several books available for use in introductory classes for beginning engineering students.

H. Queisser, *The Conquest of the Microchip* (Cambridge: Harvard University Press, 1988). Details on science and business aspects of an important technological development.

3. The Origin of Problems

1. R. Meehan, *The Atom and the Fault* (Cambridge: MIT Press, 1984). This and the following book are stories drawn on the author's career as a consulting engineer and give an excellent feeling for some of the constraints on and motivations of engineers.
2. R. Meehan, *Getting Sued and Other Tales of the Engineering Life* (Cambridge: MIT Press, 1984). See note 1.
3. J. Adams, *Conceptual Blockbusting*, 3rd. ed. (Reading, MA: Addison Wesley, 1986). A book on creativity that I wrote because of my experiences working with engineers who occasionally suffer from intellectual myopia.
4. L. White, *Medieval Technology and Social Change* (New York: Oxford University Press, 1962).
5. N. Perrin, *Giving Up the Gun* (Boston: David R. Godine, 1979). A charming little book about Japan's refusal to use firearms for a great many years.
6. N. Rosenberg, *Inside the Black Box: Technology and Economics* (New York: Cambridge University Press, 1986). An excellent study of technology by an economist who also happens to be a historian.
7. J. Burke, *Connections* (Boston: Little, Brown and Co., 1978). The book that accompanied the very successful and popular television series of the same name.
8. A. Pacey, *The Culture of Technology* (Cambridge: MIT Press, 1984). A provocative discussion of cultural forces upon technology and vice versa.

4. Design and Invention

1. C. Singer, E. Holmyard, A. Hall, T. Williams, *A History of Technology*, vol. 3 (New York: Oxford University Press, 1958).

Other Books of Interest

T. Kidder, *The Soul of a New Machine* (New York: Avon Books, 1981). A fine story of the development of a new computer by Data General Corporation—in my opinion, a must-read.

E. Morison, *From Know-How to Nowhere* (New York: Basic Books, 1974). A provocative treatment of several developments in engineering technology with emphasis on nontechnical factors.

5. Mathematics

1. J. Reuyl and R. Schutt, "Simulation of the energy performance of a solar photovoltaic residence and hybrid electric automobile in Fresno, California," *Sandia National Laboratory Report SAND81-7044* (Albuquerque, NM: 1982).
2. P. Davis and R. Hersch, *The Mathematical Experience* (Boston: Houghton Mifflin Co., 1981). An outstanding book for those interested in the nature of mathematics. It is full of fascinating information and examples, and is very well written.
3. S. Ulam, *Adventures of a Mathematician* (New York: Scribners & Sons, 1976).
4. J. Dodds and J. Moore, *Building the Wooden Sailing Ships* (New York: Facts on File Publications, 1984). Not a math book—rather a fascinating account of how wooden sailing ships were built.

Other Books of Interest

J. Gleick, *Chaos: Making a New Science* (New York: Penguin Books, 1987). A best-seller about a new field of mathematics.

R. Rucker, *Infinity and the Mind* (Cambridge: Birkhauser, 1982).

J. R. Newman, ed., *The World of Mathematics*, 4 vols. (New York: Tempus Books, 1987). This series and the next are full of information on mathematics through the ages.

M. Kline, *Mathematical Thought from Ancient to Modern Times*, 3 vols. (New York: Oxford University Press, 1990).

H. O. Peitgen and P. H. Richeter, *Beauty of Fractals* (Berlin: Springer-Verlag, 1988). Contains fascinating pictures as well as interesting applications.

D. J. Albers and G. L. Alexanderson, *Mathematical People* (New York: Contemporary Books, 1990). Well-done interviews with mathematicians and others involved with mathematics.

6. Science and Research

1. S. Richards, *Philosophy and Sociology of Science* (London: Basil Blackwell, 1987). A well-written general book about the nature of science.
2. T. Kuhn, *The Structure of Scientific Revolutions* (Chicago: University of Chicago Press, 1962). A modern classic that changed people's thinking about science.
3. D. de Solla Price, *Big Science, Little Science* (New York: Columbia University Press, 1963). A study of changes in science.
4. Richards, *Philosophy and Sociology of Science*.
5. *The Measurement of Scientific and Technical Activities* (Washington, D.C.: Organization for Economic Cooperation and Development, 1970).
6. G. Moore, "VLSI: What does the future hold?" *Electronics Australia* 42 (1980): 14.
7. R. N. Noyce, "Microelectronics," in T. Forester, ed., *The Microelectronics Revolution* (Cambridge: MIT Press, 1981).
8. *1987 Electronic Market Data Book* (Washington, D.C.: Electronic Industries Association, 1987).
9. A. L. Robinson, "Electronics and employment: displacement effect," in T. Forester, ed., *The Microelectronic Revolution* (Cambridge: MIT Press, 1981), p. 318.
10. J. Watson, *The Double Helix*, ed. G. Stent (New York: Norton, 1980). An entertaining and sometimes controversial story that gives an excellent insight into the lives and motivations of young scientists.

Other Books of Interest

R. Feynman, *Surely You're Joking Mr. Feynman* (New York: Norton, 1985). An amusing autobiography of a very famous physicist.

S. Weinberg, *The First Three Minutes* (New York: Basic Books, 1977). The weird universe just after the big bang.

R. Rhodes, *The Making of the Atomic Bomb* (New York: Simon and Schuster, 1986). A wonderful story of the people and the project.

R. March, *Physics for Poets* (New York: Contemporary Books, 1983). A standard textbook for college physics courses that are oriented toward nonmajors.

C. Djerassi, *Cantor's Dilemma* (New York: Penguin Books, 1989). A novel about a research chemist and a problem of ethics.

J. Goldberg, *Anatomy of a Scientific Discovery* (New York: Bantam, 1989). Insight into the scientific process.

Five excellent writers about science who have written many extremely readable books are Jeremy Bernstein (physics), Freeman Dyson (physics), Stephen Jay Gould (biology), Alan Lightman (cosmology), and Paul Ehrlich (population biology). Browse through some of theirs.

7. Development, Test, and Failure

1. T. Kidder, *Soul of a New Machine* (New York: Avon Books, 1981).
2. "The shuttle findings," *New York Times*, June 10, 1986, p. C11. Also the *Final Report of the Presidential Commission on the Space Shuttle Accident* (chaired by William Rogers), June 6, 1986.
3. H. Petrosky, *To Engineer Is Human* (New York: St. Martins Press, 1985). An excellent treatment of major failures.
4. See *New York Times*, Dec. 17, 1983, p. 41 of sec. 1, and Jan. 8, 1987, p. 12 of sec. 2.

Other Books of Interest

The final report of the Special Inquiry Group, Three Mile Island, chaired by Mitchell Rogovin. Nuclear Regulatory Commission (NUREG CR-1250. This and the *Challenger* reports are beautifully done and offer very detailed insight into these two well-publicized failures.

C. Perrow, *Normal Accidents* (New York: Basic Books, 1984). A social science look at technological failures.

D. Kahneman, P. Slovic, and A. Tversky, *Judgement under Uncertainty* (Cambridge: Cambridge University Press, 1982). A technical look at the way humans respond in complex situations, which is most useful in better understanding failure.

8. Manufacturing and Assembly

1. S. Miell, *History of the British Chemical Industry*, 1634–1928 (London: Benn, 1931).
2. "Allen-Bradley's stark vision," *New York Times*, Oct. 6, 1986. S. Goldstein and J. Klein, Harvard Business School case 9-687-073, 1987.

Other Books of Interest

J. P. Womack, D. T. Jones, and D. Roos, *The Machine that Changed the World* (New York: Rawson Associates, 1990). An updated story on Japanese and U.S. automobile manufacturing and production in general.

E. M. Goldratt and J. Cox, *The Goal* (Croton on Hudson, NY: North River Press, 1986). Basic truths about production presented in the form of a novel.

9. Money and Business

1. The Bob Knowlton story was originally written as a teaching case by A. Bavelas and G. Strauss and is available as a film from Round Table Film Productions.
2. Myers Briggs Type Inventory, available to qualified people from Consulting Psychologists' Press, Palo Alto, CA. Abstracted in D. Kiersy and M. Bates, *Please Understand Me* (Buffalo: Prometheus, 1978).
3. J. Friar and M. Horwitch, "The emergence of technology strategy," *Technology in Society*, 7 (1985): 143–178.

Other Books of Interest

N. Rosenberg and L. E. Birdzell, Jr., *How the West Grew Rich* (New York: Basic Books, 1986). The interaction of economics and technology in the Western world beginning with medieval times.

P. Freiberger and M. Swaine, *Fire in the Valley* (New York: McGraw Hill, 1984). One of several books chronicling the growth and success of Silicon Valley.

D. P. Levin, *Irreconcilable Differences* (New York: Penguin Books, 1990). The high-level disagreement between H. Ross Perot and General Motors.

10. Regulation

1. J. G. Burke, "Bursting boilers and federal power," *Technology and Culture*, vol. 7, no. 1 (Winter 1966): 18.
2. C. Starr, "Benefit-cost studies in socio-technical systems," from *Perspectives on Benefit—Risk Decision Making* (Washington, DC: The National Academy of Engineering, 1972).

3. R. M. Anderson, R. Perrucci, D. E. Schendel, and L. E. Trachtman, *Divided Loyalties* (West Lafayette, IN: Purdue University Press, 1980). A story about whistle-blowing on the BART project.
4. "Blowing the whistle on nuclear power: three leave G.E.," *New Engineer*, May 1976, p. 34.
5. Meehan, *Atom and the Fault.*
6. "An overview of the first seven program years: Apr. 1, 1980—March 1987," The Institute for Civil Justice, The RAND Corporation.

Another Book of Interest

M. Martin and R. Schinzinger, *Ethics in Engineering* (New York: McGraw Hill, 1983). A good overall read on the issue.

Acknowledgments

I would like to express my gratitude to the Alfred P. Sloan Foundation, and in particular to Samuel Goldberg, the program officer for the New Liberal Arts Program. The foundation both supported the development of a year-long general-studies course sequence in Technology, Science, and Mathematics at Stanford University, which was the source for much of the material in this book, and made it possible for me to free up enough time to reach the point of no-return in writing it. I would also like to thank my colleagues in the Engineering School and the Program in Values, Technology, Science and Society at Stanford University for the many things they have taught me—in particular Sandy Fetter and Bob Osserman, with whom I designed and originally taught the course sequence, and Barry Katz and Robert McGinn, who have happily helped broaden my engineering horizons.

I could not have written the book without lots of help from friends in various companies and laboratories, who gave me information, sent me photographs, and encouraged me. I also am grateful to the people at Harvard University Press for their enthusiasm, organization, design, and production of the book, and to Jonathan Dolger for steering me to them. Warm thanks are due to Howard Boyer, Senior Editor for Science at Harvard, for his encouragement during the early and late stages of the publication process. In particular I want to thank Susan Wallace, who edited the manuscript with a passion that both amazed and flattered me. I love to work with people who do their work well, and she is at the top of the list.

Finally, my debt to my wife, Marian, who consistently responds to my psychotic writing state and unreasonable demands for proofreading with a perfect blend of love, support, and humor. Even more, she served as my target audience from the time she responded to my first draft with "The writing is O.K., but I don't want to know about engineering."

Illustration Credits

Page

16 From Robert Mark, *Experiments in Gothic Structure* (Cambridge: MIT Press, 1982), reproduced by permission of the publisher. © 1982 by the Massachusetts Institute of Technology.

19 Bibliothèque nationale, Paris.

24 Science Museum, London.

25 Science Museum, London.

48 Reproduced by permission of General Electric Company.

54 Reproduced by permission of General Electric Company.

56 Reproduced by permission of General Electric Company.

60 Reproduced by permission of Gary Cleary and Cygnus Therapeutic Systems.

71 Reproduced by permission of Audemars Piguet.

73 Reproduced by permission of Raytheon Corporation.

88 Photograph by Clifford Boehmer. Rendering by John Hopkins. Reproduced by permission of Dirigo Design.

97 Reproduced by permission of Dennis Boyle.

98 Reproduced by permission of Dennis Boyle.

99 Reproduced by permission of Dennis Boyle.

100 Reproduced by permission of Dennis Boyle.

114 Reproduced by permission of David Hoffman.

127 Courtesy of NASA Ames Research Center.

130 Courtesy of Stanford Linear Accelerator and the U.S. Department of Energy.

140 Reproduced by permission of Hughes Danbury Optical Systems, Inc. Courtesy of Mill Lane Engineering.

145 After A. L. Robinson, "Electronics and Employment: Displacement Effects," in T. Forester, ed., *The Microelectronic Revolution* (Cambridge: MIT Press, 1981), reproduced by permission of the publisher. © 1981 by Massachusetts Institute of Technology.

151 Courtesy of Bruce Armbruster.

153 Reproduced by permission of Dennis Boyle.

155 Courtesy of NASA Ames Research Center.

160 Reproduced by permission of Jet Propulsion Laboratory.

162 Reproduced by permission of Jet Propulsion Laboratory.

163 Reproduced by permission of Jet Propulsion Laboratory.

175 After Henry Petroski, *To Engineer Is Human*, copyright © 1982, 1983, 1984, 1985 by H. Petroski. Reprinted with permission from St. Martin's Press, Inc., New York.

179 Reproduced by permission of Osterreichische Nationalbibliothek Bild-Archiv und Portrat-Sammlung.

182 Reproduced by permission of Drake Sorey.

185 Reproduced by permission of Miyano Machinery.

188 Courtesy of Ford Motor Company.

191 From S. M. Sze, *VLSI Technology* (New York: McGraw-Hill, 1983, 1988), reproduced by permission of the publisher. © Bell Telephone Laboratories, Inc.

191 From S. M. Sze, *VLSI Technology* (New York: McGraw-Hill, 1983, 1988), reproduced by permission of the publisher. © Bell Telephone Laboratories, Inc.

212 Reproduced by permission of Hewlett-Packard Corporation.

213 Reproduced by permission of Kent Bowen.

219 After John Friar and Mel Horwitch, "The Emergence of Technology Strategy," *Technology in Society*, 7 (1985): 151. Reproduced by permission of Pergamon Press.

223 Reproduced by permission of the *Boston Globe*.

229 Reproduced by permission of Runk/Schoenberger/Grant Heilman Photography, Inc.

242 Photograph by Lee Balterman, *Life Magazine*, reproduced by permission of Time, Inc. © Time Warner, Inc.

Index

Advisory Committee on Reactor Safe-
guards (ACRS), 235
Age of invention, 17–18
Agricola (George Bauer), 21
Agricultural revolution, 15, 72
Alberti, Leon Battista, *De re Aedificatoria*,
18
Algebra, 110–111
Allen-Bradley Company, 201–202
Alternative energy sources, 29, 227–230;
solar-powered automobile, 107–109; di-
versification of energy base, 227–228;
conservative reaction to, 228–230; future
necessity, 241
American Institute of Electrical Engineers,
26
American National Standards Institute
(ANSI), 222
American Railway Engineering Associa-
tion, 26
American Society of Civil Engineers, 26
American Society of Heating and Ventilat-
ing Engineers, 26
American Society of Mechanical Engi-
neers, 26
American Society of Testing Materials
(ASTM), 222
Aristarchus, 20
Aristotle, 11, 18, 22, 133–134
Arkwright, Richard, 23
Atomic Safety and Licensing Board
(ASLB), 235
Automobile design, 62–63
Avogadro, Amadeo, 22

Bacon, Francis, 134
Bavelas, A., 215
Bell, Alexander Graham, 23, 82
Benz, Karl, 82
Bhopal disaster, 176, 196
Biochemistry, 147–148
Biomedical engineering, 143, 147–148
Black box approach, 138
Boltzmann, L., 139
Boolean algebra, 113
Boulton, Matthew, 81
Boyle, Dennis, 96–100, 152

Boyle, Robert, 131, 134, 142
Boyle's law, 131
Brahe, Tycho, 20
British Institute of Civil Engineers, 26
Brunelleschi, F., 18
Burke, James, *Connections*, 72
Business: engineering and, 43, 53–55. *See
also* Money and business

Calculus, 113
Carnot, Sadi, 22, 135, 141
Carnot cycle, 135–137
Challenger disaster, 170–172, 226
Charles, J. A. C., 131
Charles' law, 131
Chemistry, 180, 193
Chernobyl accident, 176
Chinese technology, 69, 80
Civil engineering, 26
Clausius, R. J. E., 131
Code of Hammurabi, 9–10
Colt, Samuel, 178
Computers, 27–29, 156, 218; CT scanning,
47–57, 195; computer-aided design, 96;
effect on mathematics, 126–128; space-
craft design and, 128, 164; "debugging,"
164–165; in industry, 181, 184–185, 194–
195, 200–202
Copernicus, Nicolaus, 20
Cormack, Allan, 50–51
Creativity, 64, 78
Crick, Francis, 145
Crystal Palace Exhibition, 180
CT scanning (computerized tomography),
47–57

Daimler, Gottlieb, 82
da Vinci, Leonardo, 18, 72, 78
DC-10, 173
Descartes, René, 20
Design, 87–103; preliminary design, 44, 95–
98; detailed design, 44–45, 99–101; role
of communication, 93–94; specifications,
94–95; concept development, 95–98;
computer-aided design (CAD), 96; systems engineering, 103–105

de Solla Price, Derek, *Big Science, Little Science*, 142
Developing countries, 5, 244–245
Development, 45. *See also* Research and development
Domesday book, 14
Dykes Auto Manual, 245

Edison, Thomas, 78, 150
Electric Power Research Institute, 224
Electromagnetic radiation, 119–120, 157
Electronics, 27–28, 74, 96, 219. *See also* Integrated circuits
Energy crisis of 1970s, 227–230
Energy supply, 29. *See also* Alternative energy sources
Engelbach, R., 8
Engineering: etymology of term, 7; first schools of, 26; complexity of, 31–57; science and, 34, 39–41, 134–149; fields, 36; intellectual disciplines, 36–37; industrial process, 37–38; types of responsibilities, 37–38; budget and schedule constraints, 40, 62; mathematics and, 41–43; business and, 43, 53–55, 204–220; environment, 43–44; design and, 44–45, 95–98, 99–101; production, 46; research and development, 46–47; curriculum, 132, 143–144, 202–203
Engineers: personality and cognitive style, 32; types of work, 32–33; self-employment, 58; consultants, 58–59, 63–64; university environment for, 59, 61; in large organizations, 62–63; as managers, 62–63, 205–206, 214–220; apprenticeship, 204–205
Entropy, 138–139
Environmental problems, 30; disposal of products, 46; air pollution monitoring, 236; toxic waste disposal, 236–237; future of, 241, 243
Ethical concerns, 64, 132, 232–233
Euler equations, 121
Experimentation, 134, 150–164. *See also* Spacecraft design
Exxon spill, 176

Failure, 168–176; material fatigue, 169–170; *Challenger* disaster, 170–172, 226; human behavior and, 172–174; Hyatt Hotel collapse, 174–175; Hubble Telescope mirror, 175, 226; attitudes toward, 226–227
Fayol, Henry, 197
Feyerabend, Paul, 142
Fibonacci series, 122
Flying buttresses, 15–16
Ford, Henry, 82, 197
Ford Model T, 197
Fourier, Joseph, 22
Franklin, Rosalind, 145
Frontius, *De Aquis*, 13
Fulton, Robert, 221
Fust, Johann, 80

Galileo Galilei, 21
Genetic engineering, 27, 129, 145–148, 218; future of, 239–240
Gerstl, Joe, 32
Giffard, Henri, 23
Gilbert, William, 21
Gothic cathedrals, 15–17, 23
Great Pyramid, the, 8–9
Greek technology, 10–12, 13, 18
Gulick, Luther, 197
Gutenberg, Johann, 78, 80

Harvey, William, 21
Hawthorne effect, 198–199
Hawthorne experiments, 198–199
Heat engines, 136, 138–139
Heat transfer, 157–158
Herodotus, 8
Hero of Alexandria, 11, 23, 81, 204
Hewlett-Packard, 211–212
Hounsfield Godfrey, 51–52
Hubble Telescope Mirror, 175, 226
Human relations movement, 198–199
Huygens, Christian, 135
Hyatt Hotel collapse, 174–175
Hypotheses, 129, 132, 134
Hypothetico-deductive method, 141

Imhotep, 8
Industrial engineering, 167, 198

Industrial Revolution, 22–26, 30, 72, 178, 192–194
Ineni, 8
Institute for Civil Justice (Rand Corporation), 237–238
Institute of Electronic and Electrical Engineers, 232
Integrated circuits, 144–145; manufacture of, 189–192
Invention, 78–87; of printing, 18, 80; development of steam engine, 21, 23, 81, 135; as group effort, 80–82, 135; of assembly line, 82; National Inventors Council, 83; patents, 84, 86; ingredients for success, 87

Japanese technology, 63, 69, 79, 183; electronics, 74, 219; MITI, 76, 219; U.S. patents, 82–83
Jaquard loom, 194
Jet Propulsion Lab, 70, 93, 105, 128, 159–164, 205, 210, 236
Joule, James, 131
JPL. *See* Jet Propulsion Lab

Kepler, Johannes, 20
Khufu-onekh, 8
Kidder, Tracey, *Soul of a New Machine*, 164–165
Kinetic theory of gases, 131
Knowlton, Bob, 215
Kuhn, Thomas, *The Structure of Scientific Revolutions*, 141

Lagrange, Joseph, 22
Laplace, Pierre de, 22
Lavoisier, Antoine, 22, 142
Leaders for Manufacturing Program (MIT), 212
Learning curve, 207
Luddite movement, 25

Magnetic resonance imaging (MRI), 55, 57, 195
Management jobs, 62–63, 205–206, 214–220; problem-solving styles, 215–216;

two-culture problem, 217–218; long-term technology strategy, 218–220
Manufacturing and assembly: assembly line designs, 82, 192–193, 197, 200–202; history of, 177–183, 200–201; standardization, 178, 180, 197, 198; chemical industry, 180–181; computers in industry, 181, 184–185, 194–195, 200–202; production ability vs products, 181, 183; molding and casting, 186–187; forming, 187–188; joining, 188–189; use of robots, 189, 193, 198; mass production, 192, 197, 200; process industries, 193; economies of scale, 194, 207; "scientific management," 197–198; "human relations" movement, 198–200; unions, 199–200; "Just-In-Time" approach, 201
Marcus, Siegfried, 82
Market pull, 69–70, 72, 76
Martel, Charles, 68
Mathematical modeling, 156
Mathematics, 106–128; engineering and, 41–43, 115, 118–128; math blocks, 106, 129; quantitative thinking, 106–109; types used by engineers, 110–113; applied, 115; pure, 115; physical laws and, 115–123; effect of computers, 126–128
Maudsley, Henry, 23
Maxwell, James Clerk, 22, 119
Maxwell's equations, 120–123
McCormick, Cyrus Hall, 78, 82
Meehan, Richard: *The Atom and the Fault*, 58; *Getting Sued and Other Tales of the Engineering Life*, 59
Miall, Stephen, *History of the British Chemical Industry, 1634–1928*, 180
Military specification system (Milspecs), 224
Military technology, 10, 26–27; invention of stirrup, 14–15, 67–69; support of research, 47, 68, 74–76, 209–210; emotion as motivator, 71. *See also* Weapons engineering
Ministry of International Trade and Industry (Japan), 76
MITI. *See* Ministry of International Trade and Industry
Myers Briggs Type Inventory, 216

National Full-Scale Aerodynamics Complex, 154–155
National Inventors Council, 83
Navier-Stokes equations, 121
Newcomen, Thomas, 23, 81, 135
Newton, Isaac, 20–21, 134
Newton's laws of motion, 131; Newton's second law, 115, 117, 118, 123, 125
Nuclear power, in California, 233–235
Nuclear Regulatory Commission (NRC), 235
Nuclear weapons, 5–6

Ohm's law, 118
Operations research, 105, 157, 200
O-rings, 170–172

Pacey, Arnold, *The Culture of Technology*, 76
Papin, Denis, 81
Pascal, Blaise, 81
Paxton, Joseph, 180
Perrucci, Robert, 32
Physical laws, 115–123
Plato, 20, 133
Pollio, Marcus Vitruvius, *De Architectura*, 13, 18, 21
Polo, Marco, 80
Popper, Karl, 141
Porta, Giambaptista, 21
Positron Emission Tomography (PET), 57
Printing, 18, 80
Ptolemy, 18, 20

Quality circles, 201
Quality control, 167, 200, 202

Radiological Society of America (RSNA), 53, 55
Radon's theorem, 50–51, 123
R&D. *See* Research and development
Research and development, 46–47, 84–85, 129, 153–154, 218; military research, 47, 209–210
Reuyl, John (Energy Self-Reliance, Inc.), 108

Richards, Stewart, *Philosophy and Sociology of Science*, 142
Robots, 189, 193, 198
Roentgen, Wilhelm, 49
Roman technology, 12–14, 21–22
Root, Elisha, 178
Rosenberg, Nathan, *Inside the Black Box: Technology and Economics*, 69, 72

Savery, Thomas, 21, 81, 135
Schaeffer, Peter, 80
Science: Greek natural philosophy, 10–12, 13, 18, 133; empirical approach, 20–21; Scientific Revolution, 20–21, 134; convergence with technology, 20–22, 30, 134–149; and engineering, 34, 39–41, 134–149; and research, 129–149
Scientific knowledge, 129, 132; hypotheses, 129, 132, 134; laws, 129, 131; theories, 131–132
Scientific management, 197–198
Scientific method, 40, 135
Scientific process, 132, 141, 142
Scientific Revolution, 20–21, 134
Semiconductor, 144, 218–219
Senmut, 8
Smeaton, John, 81
Snow, C. P., 217
Society of Naval Architects and Marine Engineers, 26
Spacecraft design, 128, 158–164, 205; model of spacecraft, 159; space simulator, 159–160; Galileo spacecraft, 161–164
Space race, 74
Starr, Chauncey, 224–225
Statistical Abstract of the United States, 1988, 30, 206
Statistics, 113
Steam engine: Savery, 21, 81, 135; Newcomen, 23, 81, 135; Watt, 23, 81, 135, 221; background inventions, 81, 135; Boulton, 81; Smeaton, 81; thermodynamics and, 135–137; Industrial Revolution, 178; exploding steamship boilers, 221–222, 233
Stirrup, invention of, 14–15, 67–69
Strategic Defense Initiative (Star Wars), 45–46, 70

Strauss, G., 215
Systems engineering, 103–105; operations research, 105, 157, 200

Taylor, Frederick, 197–198
Technological determinism, 67–70, 76–77
Technological imperative, 67
Technology: dangers of, 5–6, 25, 43–44, 224; history of, 5–30; etymology of term, 7; military, 10, 14–15, 26–27, 47, 67–69, 71, 74–76, 209–210; convergence with science, 20–22, 30, 134–149; war and, 26–27, 45–46; government role in, 66–68, 71, 74, 209–210
Technology management, 77
Technology planning, 77
Technology policy, 77
Testing: of a spacecraft, 164; flight-testing aircraft, 165–169; environmental testing, 166; life testing, 166–167; quality control, 167, 200, 202
Thermodynamics, 123, 135; first law, 119, 136; second law, 119, 138; steam engine and, 135, 137; Carnot cycle, 135–137; efficiency, 136; heat engines, 136; entropy, 138–139
Thomson, J. J., 49
Three Mile Island, 176, 195, 226
Torricelli, Evangelista, 21, 39, 81, 135

Toxic waste disposal, 236–237
Transistor, 144
Trevithick, Richard, 135

Ulam, Stanislaw, *Adventures of a Mathematician*, 113
Uni, 8
Urwick, Lyndall, 197

Vesalius, Andreas, 21
Vietnam War, 68, 226
VLSI (Very Large Scale Integration) project, 218–219
von Guericke, Otto, 21, 81, 135

Watson, James, 145
Watt, James, 23, 78, 81, 135, 221
Weapons engineering, 5–6, 10, 14–15, 17–18, 66, 103. *See also* Military technology
Weapon systems, 103
Western Electric Co. (Hawthorne works), 198–199
Whewell, William, 141
Whistle blowing, 232–233
White, Lynn, *Medieval Technology and Social Change*, 67
Whitney, Eli, 78, 178
Wilkins, Maurice, 145